COCKPIT COMPUTERS

Paul Garrison

Multiengine and instrument-rated licensed pilot
Contributing editor to *Aero and Aviation Convention News*
Former managing editor of *Business and Commercial Aviation*

McGraw-Hill Book Company

New York   St. Louis   San Francisco   Auckland   Bogotá
Hamburg   Johannesburg   London   Madrid   Mexico
Montreal   New Delhi   Panama   Paris   São Paulo
Singapore   Sydney   Tokyo   Toronto

McGraw-Hill Series in Aviation

David B. Thurston, Consulting Editor

Garrison *Cockpit Computers* (1982)
Ramsey *Budget Flying* (1981)
Sacchi *Ocean Flying* (1979)
Smith *Aircraft Piston Engines* (1981)
Thomas *Personal Aircraft Maintenance* (1981)
Thurston *Design for Flying* (1978)
Thurston *Design for Safety* (1980)
Thurston *Homebuilt Aircraft* (1982)
Whempner *Corporate Aviation* (1982)

Library of Congress Cataloging in Publication Data

Garrison, Paul.
   Cockpit computers.

   (McGraw-Hill series in aviation)
   Includes index.
   1. Airplanes—Piloting—Data processing.
2. Navigation computer (Aeronautical instrument)
3. Microcomputers. I. Title. II. Series.
TL589.2.N3G37   629.132′52′02854   81-8379

ISBN 0-07-022893-0   AACR2

1 2 3 4 5 6 7 8 9 0 KPKP 8 9 8 7 6 5 4 3 2 1

The editors for this book were Jeremy Robinson and Charles P.
Ray; the designer was Elliott Epstein, and the production
supervisor was Paul Malchow. It was set in Electra by Bi-Comp,
Incorporated.

Printed and bound by The Kingsport Press.

CONTENTS

PREFACE

Long before the electronic computer entered the scene and became a virtually unavoidable part of nearly everyone's life pilots used a simple device to determine such data as the amount of fuel needed to get from point A to point B, the time it will take to get there at a given airspeed under the prevailing wind conditions, and similar flight-related information.

This device has always been called a computer, though it was, in fact, a circular slide rule which, under the best of circumstances, could only be expected to provide approximations. Millions of these computers were manufactured and sold and many are still in daily use today.

But then came the electronic computer and it was only logical that its nearly limitless capabilities should be applied to the many mathematical problems which are an integral part of today's need for precision in flying.

Two basic types of computers are in use today, providing valuable assistance to pilots in the areas of flight planning, navigation, and fuel conservation. One is the hand-held portable computer/calculator, containing large numbers of prestored aviation programs. Despite their small size, some fitting comfortably into shirt pockets, they have been developed to a high degree of sophistication. The other type is the airborne navigation computer which is an integral part of many of the more elaborate on-board navigation systems.

With the rapid proliferation of such computers it is becoming increasingly important that pilots of all categories, private, commercial, and professional, become familiar with the basic principles of computer operation so that they become more able to make full use of the computer's capabilities. This book serves to familiarize all types of pilots with current aviation computer equipment from the hand-held aviation calculators to the more complex portable aviation computers and to illustrate the procedures for the aviation calculations that each computer provides for.

I want to express my appreciation to the following companies for their cooperation in providing me with the use of their computers:

Heath Company

Hewlett-Packard

Jeppesen-Sanderson

Specialized Electronics

Texas Instruments

All tables and illustrations not otherwise credited are by the author.

Paul Garrison

1

INTRODUCTION

When we talk about the age we live in, we often like to think of it as the atomic age or the age of space exploration, but what we are really experiencing is the age of the computer. Without computers there could have been no flight to the moon, and there could be no nuclear science as we know it today.

At first, when computers became something that people would talk about at cocktail parties, we tended to be fascinated by the incredible capabilities which were ascribed to them. Then, slowly, these capabilities, largely through articles and motion pictures written by authors with more imagination than knowledge, became exaggerated to the point where we caught ourselves thinking of computers as some sort of "superhuman" beings capable of composing music or writing poetry. And by now we have become just a bit afraid of them, afraid of the day when each and every one of us has been reduced to a series of digits stored in the bottomless pit of some computer's memory, to be retrieved at will in order to invade our privacy.

The fact is that there is nothing superhuman about a computer. It is simply a machine which can do absolutely nothing without guidance by a human operator: a tool to be used to save time in accomplishing a given task. It is, in fact, nothing more than a sophisticated calculator with the added capability to permanently store given bits of information, and to retrieve this information upon request.

Any computer, no matter how "smart," can function only after it has been programmed by a human programmer. And the information it provides is meaningless unless the operator knows how to ask the right questions and is capable of correctly interpreting the answers.

In the earlier days computers were huge machines located in vast specially constructed buildings in which temperatures had to be maintained at a constant level in order to assure proper operation. They contained thousands upon thousands of vacuum tubes which required endless maintenance, and banks of magnetic tape equipment were required to perform the tasks of information storage and retrieval.

But then the Bell Laboratories invented the transistor and the term *solid state* came into existence, rapidly leading to the concept of miniaturization and then microminiaturization. The culmination of this trend toward smallness was the development of the microprocessor which made it possible to combine the capabilities of a basement full of machinery into a single computer the size of the average cigarette package.

Today, portable computers (and their poor relations, the portable electronic calculators) have become so much a part of our everyday lives that most of us would find it difficult to function without them. We can still add 2 and 2 in our heads, but when it comes to checking our bank balance or figuring out whether the "large economy size" is actually a better deal than two of the smaller packages, we automatically reach for those electronic gadgets which can do the job for us with incredible precision in the blink of an eye.

Aviation has always involved the need to solve a variety of mathematical problems and today, with the aviation environment becoming increasingly complicated, and with aircraft getting more and more sophisticated, the number of mathematical formulas affecting precision flight has grown considerably. There are still many pilots, especially those of us who have been flying for decades, who feel that approximations and guesswork are quite adequate for safe flight. And to an extent they are right. But this type of rather sloppy flying is becoming less and less acceptable in the age of constantly rising fuel prices. Fuel, once something we rarely took into serious consideration, has become such an important factor that careful planning and precision execution of any flight is an absolute prerequisite for acceptable economy.

It is little wonder, therefore, that the computer with its ability to quickly provide precise answers to virtually any aviation-related question, is becoming an ever more popular tool for the pilot, whether a 10-hour student or a 20,000-hour airline captain.

In this book we will take a close look at all the portable aviation computers and calculators which are available today and which are of some use to the pilot. As we will see, some are more useful than others, and while the operation of certain models can be learned in a matter of minutes, others require considerable study in order for the pilot to be able to make full use of all of their capabilities.

The purpose of the book is quite simple. Computers are marvelous machines, but unless the operator understands fully how they function, their usefulness is drastically reduced. In order to obtain the solution to a problem, all computers must be given certain information, often in a predetermined sequence, and then the operator must know how to interpret the results. Computers, containing no movable parts, are virtually trouble free, and their answers can be relied upon to be correct and precise, assuming that the information provided has been accurate in the first place. There is a unique acronym in computer-operator language that should always be remembered by anyone using a computer: GIGO. It stands for *Garbage In*, *Garbage Out*. What that means is that if the information provided contains an error, then the result will be equally erroneous.

PORTABLE AVIATION COMPUTERS AND CALCULATORS

2
CALCULATORS, COMPUTERS, AND LOGIC

Before getting down to dealing specifically with the different products, let's talk for a moment about certain concepts and expressions which will appear and reappear with a degree of frequency. A clear understanding of their meanings is important.

Calculator An electronic calculator is, for all practical purposes, an adding machine which, depending on its sophistication (and cost), can add, divide, subtract, and multiply, and which can also perform a considerable variety of higher mathematical functions. Many such calculators include one or several memories, meaning that we can store the result of one calculation while making another, and then recall that result to be used in further calculations. But that is all a calculator can do. It cannot automatically perform a series of mathematical steps. Each and every step involved in a mathematical calculation must be entered in the correct sequence by the operator. This means that for any purpose other than the most basic calculations, the operator must be enough of a mathematician to be familiar with such terms as *square root*, *tangent*, *sine*, *cosine*, and so on. Since most pilots, myself included, have not necessarily been trained in higher mathematics, the use of an ordinary calculator for many of the more complicated aviation problems is impractical. Even such basic problems as figuring the true airspeed (TAS) based on known indicated airspeed (IAS), temperature, and altitude, or the density altitude based on known pressure altitude and temperature, involve mathematical formulas which few of us would be able to remember whenever the solution to such a problem is called for.

Computer   To put it in the simplest of terms, a computer is a calculator which has been programmed (or can be programmed) to perform certain mathematical calculations automatically and speedily, requiring of the operator only that it be fed certain variable data which affect the outcome of the computation. Thus, using the previously mentioned problems, a computer, when "told" the IAS, altitude, and temperature, will automatically provide the value representing the TAS; or, when told the pressure altitude and temperature, will quickly come back with the figure representing the density altitude. It does this because it has previously been programmed by someone familiar with these mathematical formulas to perform the mathematical computation which is called for without further instruction from the operator. In order for it to perform its task, all the operator must do is to "tell" the computer (by using certain keys) what type of problem is to be solved, and then to feed in (again using usually numerical keys) the variable data which is known and which the computer must use in its computation.

All computers consist primarily of a keyboard and a display window. The keyboard is the means by which the operator communicates with the computer. Each key has one or several specific functions and, when pressed, it tells the computer what the operator wants it to know. The display window serves two functions. While information is fed into the computer by the operator, it will display the information being entered, thus providing the operator with an opportunity to check this data and to make certain that no mistake has been made. Then, once the computation has been performed, the display window will show the result of the computation. Some computers can be attached to printers which will print anything that appears in the display window, thus providing the operator with a so-called *hard copy* of the information fed into the computer and the eventual result of the computation.

In computer terminology, the machines themselves are referred to as *hardware*, while the programs either already stored in the computer, or available for use by the computer, are referred to as *software*. All printed material provided by computer-activated printers is referred to as *hard copy*.

Logic   The computers which we will be discussing in this book operate in accordance with two different types of logic, meaning, from the operator's point of view, the sequence in which data are entered into the computer. These two types of logic are referred to as *arithmetic logic* and *Polish logic*. Arithmetic logic means that data are entered into the computer in the same consecutive order as would be used if the problem were figured out with pencil and paper. Thus, in order to add 2 and 2 we press the following keys in logical succession: **2 + 2 =** and the computer (or calculator) will show *4*.

Polish logic (credited to the Polish logician J. Lukasieqicz) requires a different sequential order of entering information. In this case, in order to perform the same calculation, we press the appropriate keys as follows: **2 ENT 2 +** and again the display window will show *4*. In other words, we first enter a value by pressing the appropriate numerical keys, then we press **ENT** to enter that value, then we enter a second value, and then, by again pressing the right key, we tell the computer what to do with that second value in relation to the first. This would, at first, seem to be a convoluted way of doing things, but it is actually quite easy to get used to. The reason for using Polish logic is that it permits greater versatility of the computer or calculator.

Because for many decades it has been referred to as a computer, we will also take a brief look at the aviation computer that is generally known as the E6b (a designation which is a hangover from the military). This is, of course, not a computer at all. Technically it is a circular slide rule in which data can be placed in a given relation to one another, providing an approximation of the correct answer to a mathematical problem. It is important to remember that all answers obtained through the use of the E6b are strictly approximations, precision being limited by the size of the unit. Large E6bs do a better job than small ones, but none can approach the kind of precision which is available by using their electronic counterparts.

3

THE E6b PROFESSIONAL FLIGHT COMPUTER

Despite its generally accepted definition, the E6b is not a computer but rather a slide rule (see Figure 3-1). It consists of a fixed base and a circular portion which can be rotated. Around that circular portion, on the fixed base, are a bunch of numbers which can, depending on the problem being worked, represent miles (either nautical or statute), TAS, fuel, or minutes. The circular rotating center section has an outer and an inner scale around its periphery. The outer scale can represent either minutes or calibrated airspeed, and the inner scale is marked in hours and minutes. Problems are solved by rotating the circular portion and producing two sets of relationships between the data on the scale on the fixed base and those on either one of the two scales on the movable portion.

Furthermore the rotating portion has a window under which additional data are printed on the fixed base. This window is used in determining TAS and density altitude.

In addition many models have a temperature conversion scale printed somewhere on the fixed base, and most include two marks which permit easy conversion of statute miles to nautical miles and vice versa.

The more expensive versions of the E6b include a similar arrangement on the backside of the computer, used to determine heading and ground speed under given wind conditions (see Figure 3-2). It, too, consists of a fixed base plus a circular rotating section, this latter being transparent in its center portion. In addition there is a slide which can be moved up and down and which is needed in order to solve these wind-related problems. With few exceptions, the only time pilots use this portion of the E6b is when preparing for and taking written tests. The routine is simply too complicated and relatively inaccurate to be of much use in day-to-day operations.

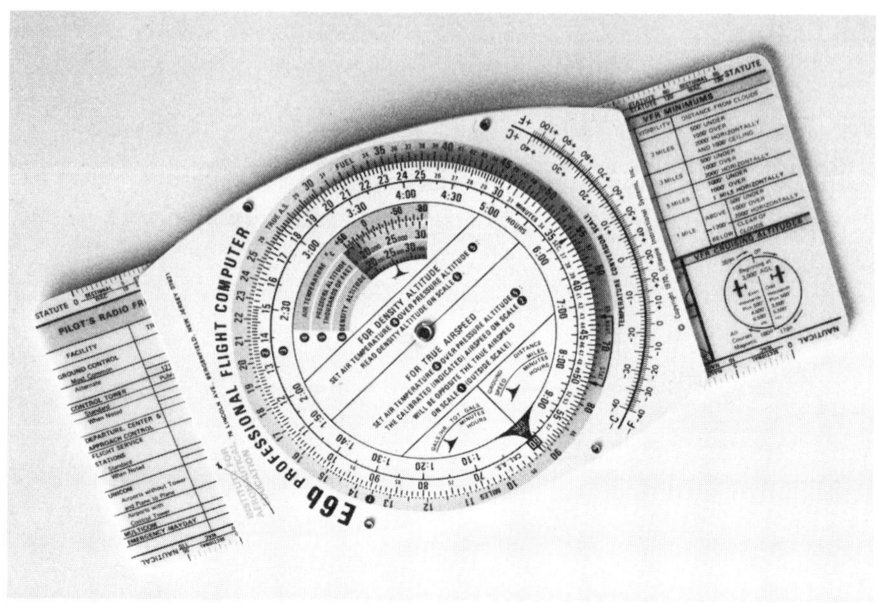

FIGURE 3-1   The E6b Professional Flight Computer.

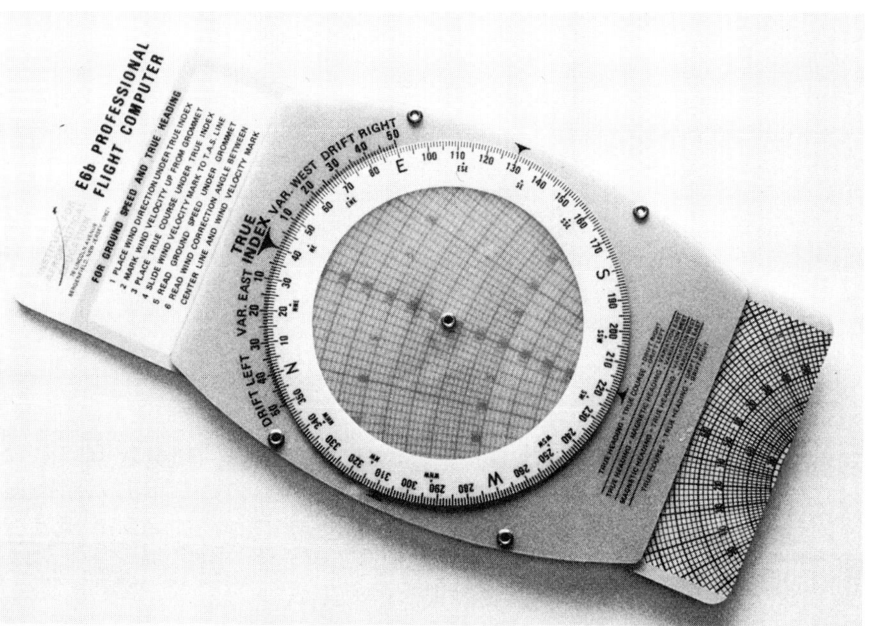

FIGURE 3-2   The backside of the E6b, designed to be used in solving wind-related problems.

Here is what it involves: To determine ground speed and heading based on known TAS, wind components, and course, first place the wind direction under the **TRUE INDEX** mark. Now take a pencil and make a dot on the transparent window, marking the wind velocity measured up from the center grommet. Rotate the circular portion and place the true course under the **TRUE INDEX** mark. Now slide the movable slide up or down until that pencil dot (representing wind velocity) coincides with the line on the slide which represents the TAS. At this point we can read the ground speed under the center grommet, and we can find the wind-correction angle between the centerline on the slide and the pencil dot.

Not too long ago a company called Walnut Products introduced a redesigned version of the E6b which is being marketed under the name CalculatorPLUS. Like the E6b, it consists of a fixed base and some rotating sections which, when placed into the appropriate relationship, will provide approximate answers to a number of aviation problems. It can be used in working fuel-, distance-, and time-related problems, to determine density altitude and TAS, to predict temperatures at altitude and freezing levels, to solve wind triangles and other wind-related problems, to determine traffic pattern entry and headings, to find holding-pattern entries, automatic direction finder (ADF) headings, very high frequency omnidirectional range (VOR) interception angles, distances from VORs, and bearings to or from landmarks.

Personally, I have found it difficult to use, but it is possible that once a pilot becomes completely familiar with its workings, it'll do a good job for him or her.

4

NATIONAL SEMICONDUCTOR'S MATHEMATICIAN

This is an ordinary electronic calculator, operating either on a 9-volt battery or on house current via a converter (see Figure 4-1). The reason it is included in the collection of portable computers (even though it isn't a computer) is that many of the basic aviation problems can be solved by using any one of the ordinary calculators, and I believe that the comparison will be worth while for those who rarely if ever need to bother with the more advanced and difficult problems which such calculators are incapable of handling.

This particular calculator consists of a 32-key keyboard and a display window which accommodates eight digits plus a decimal point. Nine of the keys are given two designations, meaning that they are capable of performing two different functions, depending on whether the function key (**F**) is pressed prior to the use of any of these keys. Only someone thoroughly conversant in the field of higher mathematics will be able to take full advantage of all the capabilities of this calculator. In using it with reference to aviation problems we will, therefore, ignore all those special function keys and simply use it as an ordinary low-cost model.

The keys we are concerned with are those marked **O** through **9**, the decimal point, a key marked **C**, the ÷, ×, −, and + keys, and two keys marked **MS** and **MR**, which stand for *memory storage* and *memory recall*, providing access to the one-memory capability. And then there is the **ENT** key which enters any displayed value into the calculator. The only key which is not self-explanatory is the **C** key which, when pressed, removes the last entry from the display and returns the next-to-last entry to the display.

This particular calculator operates on the principle of Polish logic,

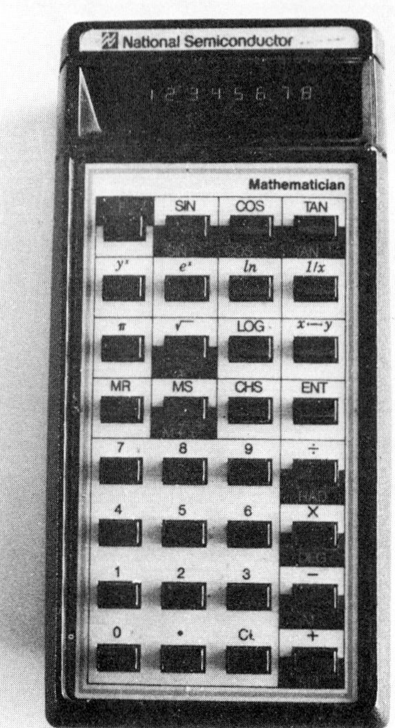

FIGURE 4-1 National Semiconductor's Mathematician is an ordinary electronic calculator with one memory. It functions on the principle of Polish logic.

which is the reason that there is no **=** key. To perform a calculation the first value is entered, then the second value is pressed, followed by the appropriate function key, which will cause the result to be displayed: **50 ENT 50 +** and the display now shows *100*.

This calculator, like all calculators, cannot deal with values expressed in hours:minutes:seconds. The only manner in which time can be used in calculations is by converting it to decimal values. The formula for converting minutes or seconds to decimals of hours or minutes works like this: Divide the number of minutes (or seconds) by 60 and the result will be the appropriate decimal (30 ÷ 60 = 0.5). Conversely, to change decimal time values into hours:minutes:seconds, multiply the decimal value by 60 (0.5 × 60 = 30).

The primary difference between using a calculator or a computer to solve aviation-related problems is that the calculator requires that the operator know the mathematical formulas involved, while the computers have these formulas already stored as part of the standard program.

5

TEXAS INSTRUMENTS' BASIC AVIATION COMPUTER

Texas Instruments, in cooperation with Jeppesen-Sanderson, developed a small and relatively inexpensive aviation computer which is being marketed by several different companies under a variety of names (see Figure 5-1). Jeppesen sells it as the Avstar, Cessna calls it the Sky/Comp, Beech is selling it under the name of Navstar, and Piper and Hughes Helicopters call it the Avstar. They are all identical, all containing the same number and type of aviation programs which are capable of solving a fairly large number of the more basic aviation-related problems.

The computer measures 3×5.25 inches, is 0.5 inches thick, and weighs about 4 ounces. The display uses liquid crystal data which requires a minimum of electrical current, resulting in a battery life of up to 1000 hours of operation. The keyboard consists of 40 keys, many of which perform two different functions, depending on the mode of operation. The display window can accommodate eight digits plus a decimal point, with the last digit automatically rounding off subsequent fractions to the nearest higher or lower number, fractions from 1 to 4 being rounded off to the lower number, and 5 to 9 being rounded off to the higher number. The computer automatically shuts itself off after some 5 minutes during which no entry has been made, and when it does, all previously displayed or entered information is lost. To cancel this automatic shutoff feature, the operator has to press three keys (**0 +/− =**) and the information and display will be retained until the computer is turned off by the operator. The computer sells for just under $70 in 1981 $'s.

The 40 keys and their designations are shown in Figure 5-1. They are divided into groups, outlined in black, red, and blue, plus other

FIGURE 5-1 Texas Instruments in cooperation with Jeppesen-Sanderson has developed an electronic replacement for the E6b, the Avstar.

keys which fall outside those outlined areas. It is important to understand the meaning of each of those key identifications and the purpose of those outlines in order to use the computer correctly.

The calculator portion of the computer is concentrated at the bottom half of the keyboard. Here we have keys identified by the digits **1** through **0**, plus a decimal point and a change-sign key (**+/−**) in the three center rows, and a divide (**÷**), multiply (**×**), subtract (**−**), add (**+**), and equal (**=**) key on the right. These keys permit the operator to use the computer as a standard calculator (using arithmetic logic), and the digit keys are also used to enter the required information into the computer when it is used to perform aviation-related problem solving.

On the left side are four keys identified as **T:**, **Sto**, **Rcl**, and **Conv**.

The **T:** key causes data entered to appear as hours, minutes, and seconds (*00-00-00*). It works like this: If you want to enter 2 hours, 34 minutes, and 3 seconds, you press **2**, then **T:**, then **34**, then **T:**, then **03**, and the display will show *2-34-03*. It should be noted here that the computer can perform mathematical calculations with such time data. Thus, if we follow the above by pressing multiply (**×**) and then **2**, followed by **=**, the display will show *5-08-06*. But there are limits to this capability. If, for instance, you are trying to add an hour:minute:second value to a time value in terms of hours (or minutes) and decimals of hours (or minutes), the result will appear in hours (minutes) and decimals thereof. Thus 1-00-00 + 1.5 will appear as *2.5*. In other words, the computer will multiply or divide hours:minutes:seconds by values expressed in decimals, and it will do the same when division is required, in each instance producing a result expressed in hours:minutes:seconds. But it will not do it when asked to either add or subtract decimal values from (or to) hours:minutes:seconds. But the computer will convert decimal time values into hours:minutes: seconds in a very simple manner. All the operator has to do is to press three keys: **+ T: =**, and the conversion will instantly appear. Thus, say, 2.83 hours, in order to be converted into something that makes sense to us, requires the following keystrokes: **2 . 8 7 + T: =**, and the display will show *2-52-12*. Conversely, time values can be converted to decimal displays by again using three keys: **+ 0 =**. Thus to reconvert the above time value we first enter it and then initiate the conversion. The sequence of keystrokes would be: **2 T: 52 T: 12 + 0 =**, and the display will show *2.87*. In aviation, where we are frequently confronted with time figures in decimals, this conversion capability is an important feature. One more thing about time. If we happen to enter an impossible time figure in terms of hours:minutes:seconds, such as 5:78:93, the computer will accept it without argument. But anyone looking at that display would immediately know that this is foolish. To transpose it to an acceptable value, we simply press the **=** key, and the display will instantly change to *6-19-33*, which, of course, is what all those hours, minutes, and seconds add up to.

The next key on the left row below the **T:** key is one marked with **Sto**. **Sto** stands for *storage*, and it means that when we press that key, whatever value happens to be displayed in the display window at that time, will be entered into the memory, where it will remain until it is either recalled or the computer is shut off. This is a valuable capability as it will permit us to remember certain results of previous calculations or other values, while we conduct any number of additional calculations. Then, when we need to recall that value in order to use it in conjunction with the results obtained in the meantime, all we have to do is recall it.

The key below **Sto** is marked **Rcl**. That stands for *recall*, and it is used in order to recall whatever value was previously stored in the memory. Thus, if we have performed a calculation according to the result of which the fuel burned in order to get to altitude will be 3.8 gallons, we can store 3.8 in the computer memory. We can then figure out how much fuel we need for the cruise portion of the flight under prevailing wind conditions. Let's say the result of that computation shows *43.2* gallons. We now press the **+** key and recall the stored figure, then press the **=** key and the display will show *47*. We now press **Sto** again to once more store that value, and we compute the amount of fuel used during descent, say, 2.3 gallons. Then, using the same procedure to add the stored figure to that value, we recall the stored value and the result will appear as *49.3*. Since the only alternate means of remembering figures would be to use paper and pencil, and since paper and pencil are not always handy, this memory feature can prove to be quite valuable.

The bottom key in the left-hand row is marked with a *blue* **Conv**. The **Conv** stands for *conversion*, and the fact that it is blue reminds the operator that once that key has been pressed, all keys which have blue designations in addition to the primary black designations will now use the function identified in blue. In this mode the computer has been programmed to convert certain values to certain other values. It will convert U.S. (avoirdupois) pounds (**LB**) to kilograms (**KG**), and it will do the opposite. Further, it will convert feet (**FT**) to meters (**M**) and vice versa, nautical miles (**NT**) to kilometers (**KM**) and vise versa, gallons (**G**) to liters (**L**) and vice versa, degrees Fahrenheit (**°F**) to degrees Celsius (**°C**) and vice versa, and statute miles (**ST**) to nautical miles (**NT**) and vice versa. Personally, this author is puzzled by some of these designations and conversion values which have been chosen by the designers of the programs. First of all, why statute miles should be designated as ST, instead of the commonly used sm (or SM), and nautical miles shown as NT instead of nm (or NM) seems capricious at best. The other puzzle is why, instead of some of the less useful conversions, such as feet to meters, a conversion of gallons of aviation gas to pounds and gallons of jet fuel to pounds wasn't selected as something which we need much more frequently. Granted, multiplying or dividing by either 6 or 6.7 is not terribly difficult, but it somehow would have seemed to me to be a more intelligent use of the programs which were available to the designers.

On the top row of keys, on the far right is a key marked **On/C** which stands for turning the computer *on* and for *correction*. Pressing the key once will turn the computer on and the display will show *0*. Then, if we find that we have made a mistake during a calculation, it can be used to erase the faulty entry. Here is how that works: Let's say we want to add 25 plus 25 plus 3. We press **2 5 + 2 5 +**. The display will now read *50*. If now, accidentally, we press **4** instead of **3**, the display will show *4*, and we'll know that we have made a mistake. To correct it we now press the **On/C** key once and the display will replace the *4* with a *0*. We now press **3** and the **=** and the display will show *53*. A warning: If we press the **On/C** key twice, it will erase everything that has previously been used in the calculation (except anything stored in the memory) and we'll have to start from scratch. Also, if after pressing the **On/C** key once in order to make a correction, we then press the **+** (or any other function) key before entering the corrected digit(s), the computer will perform a computation using the last remembered figure. Thus, if the last remembered figure was 50 and you then pressed **+**, then **On/C**, and then **+** again, the display window will show *100*. Similarly, if that last figure was 50 and you pressed **×**, then **On/C** and **×** again (or, in fact, any other function key), the display window will show *2500* (the result of 50 × 50).

The function keys can be pressed any number of times in succession, and the computer will always perform the applicable function with the value displayed in the display window. Thus, by pressing **2** and then **+ + +** and so on, the display will show *2*, *4*, *8*, *16*, and so on. This works with all of the function keys (plus, minus, divide, multiply), and comes in handy when the same figure has to be used over and over again, such as when you're trying to figure out, when sitting in Las Vegas with your $1 on black, how often the wheel has to come up black in succession in order to cause you to win $1 million. (The answer is 20 times and your winnings would be $1,048,576.)

Next to the **On/C** key is a key marked **Off**. When it is pressed everything in the computer, including anything stored in the memory, is erased, and the computer is simply turned off.

On the top row on the far left is a key marked **Comp**. It is used after all applicable information has been entered into the computer to command it to start to compute. It must always be used in conjunction with one other key which will add to that command the type of computation which is desired. Thus, pressing **Comp** and then **D Alt** will tell the computer that it is to use the previously entered information in order to arrive at the density altitude. This is important because in many cases the same information can be used to provide values applicable to several different problems. For instance, in order to determine ground speed with reference to TAS, or heading with reference to course, the same variable data (TAS, course, wind direction, wind velocity) are required, and the computer must know which value we are after.

All the remaining keys, all of which fall into those three areas encircled by the black, red, and blue lines, are keys which are used in solving a fairly large number of the more common flight-planning and navigation problems. The three keys on the left side, marked **Alt/AS**, **Wind**, and **T-S-D**, inform the computer of the category or type of problem which we want to solve. **Alt/AS** stands for *altitude* and *airspeed*, and is used most often to determine TAS or density altitude. The next, **Wind**, designates the problem as *wind* related and it is used most to determine ground speed, heading, or wind components in cruise. The **T-S-D** key activates *time*, *speed*, and *distance* problems, such as estimated time en route (ETE) and so on.

In later chapters we will look at each of the aviation problems and show how the various types of computers arrive at the answers, and what actions are required of the operator. Suffice it to say here that once the computer has been informed of the type of program to be solved (by pressing **Alt/AS**, **Wind**, or **T-S-D** keys), the known variable data can be fed into the computer in any consecutive order (it makes no difference) as long as each digital entry is followed by pressing the appropriate key which identifies that entry for what it is. Thus, pressing **6 3 4 4**, followed by **P Alt**, tells the computer that the 6344 value represents pressure altitude.

The computer can also be used to solve weight-and-balance problems, but these use strictly the calculator function of the computer. There is no prestored weight-and-balance program.

Even though this computer is marketed under several different names, we will, from here on in, refer to it under the Avstar name; the name under which it is sold by Jeppesen-Sanderson. But no matter whether your computer was purchased from Jeppesen-Sanderson, or from Cessna, Beech, Piper, or Hughes under another name, all functions are identical.

6

THE COMMODORE
N60 NAVIGATOR

This computer which is being advertised and distributed by Carbooks of Sea Cliff, New York, sells for just under $100. At first glance it tends to be a bit confusing with its 60-key keyboard, 41 of which are designed to perform dual functions (see Figure 6-1). And studying the instruction booklet doesn't help a great deal. Not only is it confusing, but apparently it was written by someone who was uncertain about the meaning of what he (or she) was writing, using *sings* when the word should have been *signs* and in the weight-and-balance program using *movement* instead of *moment*.

But once one gets over the hurdle of trying to figure out what is meant by the various instructions, it becomes clear that the computer has certain unique capabilities. In addition to the usual calculator and higher mathematics functions (it employs arithmetic logic), it has two memories and can be used to solve most of the more common aviation problems, as is described in detail in Chapters 13 to 32. Two unique features involve a program which corrects distance measuring equipment (DME) speeds for slant range, both vertically and horizontally, and another which permits plotting area navigation (RNAV) courses based on latitude/longitude information for the departure and destination points. Both are described in detail in the applicable chapters.

But the computer also contains a number of functions which, to me, at least, appear to be completely useless. There is an OFF-COURSE CORRECTION program which, when told the distance between departure and destination, the distance flown, and the distance off course, will come up with a course correction in terms of degrees, minutes, and seconds and the distance from the current

FIGURE 6-1 The Commodore N60 Navigator is marketed as a direct-mail item by Carbooks of Sea Cliff, New York.

position to the destination. Quite frankly, if I know how many miles I have flown and how many miles I am off course, then I should be able to figure out what to do in order to get back on course without the help of a computer.

Another relatively useless program involves finding the true temperature based on pressure altitude and the indicated temperature. The need for this information is certainly not one that comes up all the time, though in order to be completely accurate in figuring TAS or density altitude, the true temperature rather than indicated temperature should be used. For some incomprehensible reason the instruction booklet asks that pressure altitude, calibrated airspeed (CAS), and indicated temperature be entered into the computer to run the program. That is patent nonsense. CAS has absolutely nothing to do with this problem and the computer figures the problem just as well if the CAS value is left out.

In addition to figuring TAS, the computer will also figure Mach numbers, which may be nice for people flying jets, but I have a feeling that most jet pilots would prefer and can afford some of the more sophisticated aviation computers.

Then there are a whole bunch of programs concerned with determining the present position based on information relative to one or two VORs, rhumb-line navigation, and so on. The trouble with all of these types of programs is that in the real world they are of no practical use. Any pilot who knows all of the information required in order to make these programs work would know how to determine his or her present position with an adequate degree of accuracy, without the help of a computer.

Using the N60 for weight-and-balance computations is made easier by the availability of two memories and the fact that values can be added to the data stored in these memories by using just one keystroke. Thus the operator is relieved of the need to recall data from the memory each time when an addition must be made, and then to restore them again into the memory. Otherwise the process is the same as with a standard calculator.

One of the most useless sections of the instruction booklet is the one entitled "Great Circle Plotting." This is not a computer program but simply a series of calculations which require in excess of 100 keystrokes. And even then it is not true great-circle plotting, but rather a means of establishing a certain latitude on a given longitude, more or less halfway between departure and destination at which the great-circle route would intercept that longitude. I tried to work this series of calculations for a flight from Chicago to Los Angeles. I tried it five times and each time the computer came up with an obviously wrong answer. More probably than not I made a mistake somewhere in those 100-plus keystrokes, but after a while I simply gave up, and I can't imagine a pilot who habitually flies distances of the length at which this type of navigation is of consequence, trying to work this cumbersome procedure; not to mention that aircraft designed for long-range travel are usually equipped with very low frequency (VLF)/Omega or inertial navigation systems (INS) which automatically compute great-circle distances on a continuing basis.

Overall, my personal feeling is that 60 keys are simply too many to avoid continuous confusion. In addition I find the appearance of the N60 to be excessively garish with its raised silver lettering and all those shiny blue, grey, and white, white-bordered keys. Still, it's pretty good, especially for those who want a combination of aviation computer and higher-mathematics calculator at a reasonable price.

7

THE HEATHKIT OC-1401 AVIATION COMPUTER

Among the multitude of electronic gimmickry which the Heathkit people manufacture and market for the intrepid do-it-yourselfer is an aviation computer which, for the rather modest price of under $150 (in 1981 $'s) is quite a remarkable machine (see Figure 7-1). It comes in the form of hundreds of mostly tiny parts, all of which must be assembled and installed by the purchaser, in accordance with detailed step-by-step instructions.

The instructions are clearly stated and relatively easy to follow even by one who does not consider him- or herself an expert in such matters. But it does require a steady hand, fairly decent eyesight, a few ordinary tools, and preferably a minimum of clumsiness. My past involvement with model railroading stood me in good stead, and it took me about one full day to accomplish the task, not counting the time wasted as a result of ruining one of the transistors, which I had to replace.

The OC-1401 consists of a ten-character display using light-emitting diodes (LED), two slide switches, one for **OFF/ON** and the other for **NORMAL/STANDBY**. This latter switch, when in the **STANDBY** mode, allows the computer to retain information stored in it for considerable periods of time with virtually no drain on the batteries. In that mode no calculations can be performed and there is no display. In order to recall the information stored in the computer or to perform additional calculations, the switch must be reset to **NORMAL**. The keyboard itself consists of 36 keys, identified in groups by easily distinguished color coding; the orange group being used primarily for preflight planning, the yellow group representing in-flight operations, and the blue group concerning navigation problems.

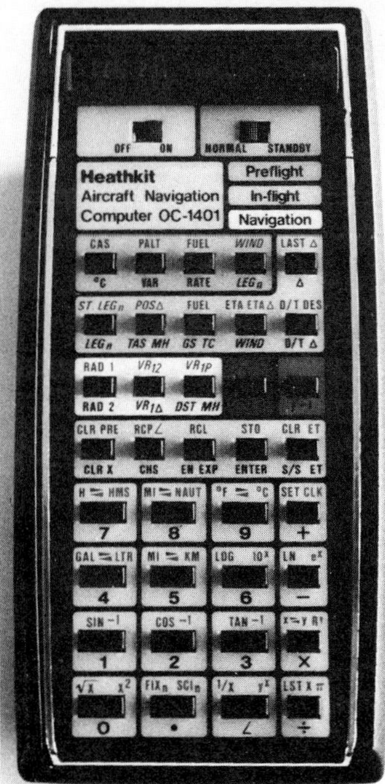

FIGURE 7-1 The Heathkit OC-1401 Aircraft Navigation Computer is sold in kit form and must be assembled by the purchaser. In this illustration the IC . . . display indicates that a mistake was made in the assembly. I had installed one diode backwards.

All keys have either two or three different functions. The functions identified in black are what might be referred to as the normal functions for each of the keys. Other functions are printed in green and brown. In order to have the keys perform the functions shown in green, the green **F** key must first be pressed. In order to cause the keys to perform the functions printed in brown, the brown **F-1** key must be used first.

The key functions are as follows:

 °C for temperature in degrees Celsius (black)

 CAS for calibrated airspeed (green)

VAR   for magnetic variation (black)

PALT   for pressure altitude (green)

RATE   for fuel consumption rate (black)

FUEL   for fuel on board (green)

LEG$_n$ for up to nine legs in vector format (black)

WIND   for wind components (green)

That takes care of the preflight keys, the left four keys on the top row. We now look at the in-flight keys:

ST.LEG$_n$ tells the computer to start the desired flight leg and to display TAS, magnetic heading (green)

LEG$_n$ recalls the nth leg previously entered (black)

POSΔ   causes updated wind vector to be displayed (green)

TAS MH   calculates TAS and magnetic heading (black)

FUEL   shows current fuel remaining (green)

GS TC   calculates ground speed and displays course (black)

ETA   displays estimated time of arrival (green)

ETAΔ   displays estimated time of arrival (ETA) at next checkpoint (brown)

WIND   recalls current wind components (black)

D/T DES   distance and time to destination (green)

D/TΔ   distance and time to next checkpoint (black)

Δ   when pressed identifies arrival at checkpoint and causes the computer to start the next leg (black)

LASTΔ   reverts all data to what they were before Δ was pressed (green)

So much for the in-flight keys. We now look at the navigation keys:

RAD 1   radial from VOR 1 (green)

RAD 2   radial from VOR 2 (black)

VR$_{12}$ distance and angle from VOR 1 to VOR 2 (green)

VR$_{1\Delta}$ distance and angle from VOR 1 to checkpoint (black)

VR₁ₚ calculates distance from VOR 1 to aircraft (green)

DIST MH distance and heading to checkpoint (black)

Also in that row of keys are the two function keys marked in green and brown (**F** and **F-1**). The next row consists of keys designed to perform a variety of functions:

CLR PRE clear previous entry (green)

CLR X clear display to *0.00* (black)

RCP ∠ reciprocal of radial, heading, course, etc. (green)

CHS change sign (black)

RCL recall from memory (green)

EN EXP exponent (black)

STO store in memory (green)

ENTER enter previously pressed digits (black)

CLR stops elapsed time, clears to zero (green)

ET displays elapsed time (brown)

S/S ET starts or stops elapsed time (black)

The next four rows of keys contain the digit keys, **1** through **0**, the function keys (**+, −, ×, ÷**), a decimal point and a key marked ∠ which stands for vector format. In addition, each of these keys has one or two additional functions:

H=HMS changes time in decimals to hours : minutes : seconds and vice versa (green and brown)

MI=NAUT converts statute miles to nautical miles and vice versa (green and brown)

°F=°C converts degrees Fahrenheit to degrees Celsius and vice versa (green and brown)

SET sets the clock to displayed time and starts it (green)

CLK accesses the true time display (brown)

GAL=LTR converts gallons to liters and vice versa (green and brown)

MI=KM converts miles to kilometers and vice versa (green and brown)

The balance of the alternate functions relate to the use of the computer for calculations using higher mathematics.

As becomes obvious when we realize the wide variety of functions offered by those 36 keys, the computer has vastly greater capabilities than we might expect from something as small and, of all things, assembled and installed by the owner. Two of these features are of primary importance. One is the fact that the computer contains a stack of nine memories, meaning that all data applicable to nine legs of a flight (or nine one-leg flights) can be stored and retained in memory. Though these memories are volatile, meaning that once the computer is shut off, all stored information is lost, the standby feature would permit, theoretically at least, turning these memories into the equivalent of nonvolatile memories. An operator could, if he or she so chose, store the applicable data for several frequently flown trips in the computer. Then, by never actually shutting it off, but by keeping it on standby when not in use, plugging it into house current or current supplied by an automobile or aircraft (using the appropriate converters which are part of the kit), and letting the nickel-cadmium (nicad) batteries do their job while no converter is connected, the operator could maintain that information in the computer for months on end.

The other unique function is the timer system. Not only does it keep track of actual time, assuming it was initially set to conform with either local or Greenwich time, it also can be called upon to display elapsed time, estimated time of arrival at either the destination or the next checkpoint at any time during the flight, and it keeps track of the amount of fuel burned and fuel remaining on a continuous basis without interrupting its primary timing function.

The OC-1401 appears to be the only one among the lower-priced computers which is capable of planning and then following through on entire flights consisting of a series of sequential legs. Like all the more sophisticated units, it takes a bit of trial and error and practice to get full use out of it. But once the operator has gotten the hang of how things work, it will do a very useful job for the active and serious pilot, especially if the aircraft involved is of relatively high performance, where precision means meaningful savings in terms of fuel and time.

To demonstrate the multileg flight-planning capability of the computer, let's plan a flight, consisting of four legs, from Albuquerque via Santa Fe, Las Vegas (New Mexico), Pueblo (Colorado), to Denver (see Figure 7-2).

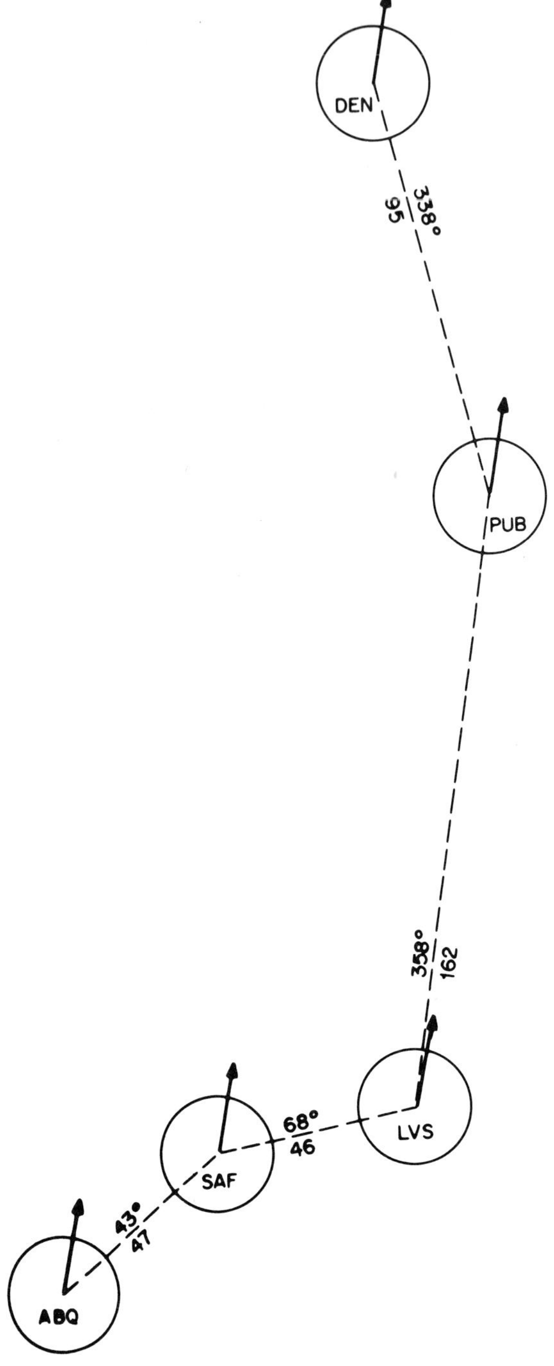

FIGURE 7-2 The four-leg flight from Albuquerque, New Mexico, to Denver, Colorado. (Not to exact scale.)

First we enter the CAS, the expected temperature at altitude, the cruising altitude, the magnetic variation, the fuel on board, and the fuel flow:

Data to be entered	Press key(s)	Display	Description of the display meaning
CAS 130 kn	1	1.	
	3	13.	
	0	130.	Calibrated airspeed
Enter	F	130.	
	CAS	130.00	
Temperature in °F	8	8.	
	5	85.	85°F
Convert to °C	F	85.	
	°F=°C	29.44	29.44°C
Enter	°C	29.44	
Cruise level	1	1.	
	1	11.	
	5	115.	
	0	1150.	
	0	11500.	11,500 feet msl cruise
Enter	F	11500.	
	PALT	11500.00	
Magnetic variation	1	1.	
	2	12.	
	CHS	−12.	Easterly variation
Enter	VAR	−12.00	
Fuel on board	8	8.	
	4	84.	
Enter	F	84.	
	FUEL	84.00	
Fuel flow in gal/h	1	1.	
	3	13.	13 gal/h
Enter	RATE	13.00	
Average forecast wind	4	4.	
	5	45.	Forecast wind velocity
	∠	45.L.	
	3	45.L3.	
	1	45.L31.	
	5	45.L315.	Forecast wind: 45 kn from 315° (true)
Enter	F	45.L315.	
	WIND	45.L315.0	
Leg 1, dist., course	4	4.	
	7	47.	
	∠	47.L.	

(Continued)

Data to be entered	Press key(s)	Display	Description of the display meaning
	4	47.L4.	
	3	47.L43.	Leg 1: 47 nmi; 043° (magnetic)
Enter	LEG$_n$	47.L43.	
	1	47.00 L4.00	
Leg 2, dist., course	4	4.	
	6	46.	
	∠	46.L.	
	6	46.L6.	
	8	46.L68.	Leg 2: 46 nmi; 068°
Enter	LEG$_n$	46.L68.	
	2	46.00 L68.00	
Leg 3, dist., course	1	1.	
	6	16.	
	2	162.	
	∠	162.L.	
	3	162.L3.	
	5	162.L35.	
	8	162.L358.	Leg 3: 162 nmi; 358°
Enter	LEG$_n$	162.L358.	
	3	162.00 L358.0	
Leg 4, dist., course	9	9.	
	5	95.	
	∠	95.L.	
	3	95.L3.	
	3	95.L33.	
	8	95.L338.	Leg 4: 95 nmi; 338°
Enter	LEG$_n$	95.L338.	
	4	95.00 L338.0	
Compute:	LEG$_n$	95.00 L338.0	
	1	47.00 L43.00	
	TAS MH	164.39 L15.12	164.39 kn; 015.12°
	GS TC	156.55 L43.00	156.55 kn; 043°
	D/T△	47.0 0.18.00	47 nmi; 0:18:00 h
	LEG$_n$	47.0 0.18.00	
	2	46.00 L68.00	
	TAS MH	164.39 L41.41	164.39 kn; 041.41°
	GS TC	176.67 L68.00	176.67 kn; 068°
	D/T	93.0 0.33.38	93 nmi; 0:33:38 h
	LEG$_n$	93.0 0.33.38	
	3	162.00 L358.0	
	TAS MH	164.39 L335.2	164.39 kn; 335.2°
	GS TC	128.59 L358.0	128.59 kn; 358°
	D/T	255. 1.49.13	255 nmi; 1:49:13 h
	LEG$_n$	255. 1.49.13	
	4	95.00 L338.0	

Data to be entered	Press key(s)	Display	Description of the display meaning
	TAS MH	*164.39 L319.9*	164.39 kn; 319.9°
	GS TC	*122.02 L338.0*	122.02 kn; 338°
	D/T	*350.00 2.35.56*	350 nmi; 2:35:56 h

As the breakdown of this four-leg flight plan shows, the computer will provide the TAS, the ground speed and the magnetic heading for each leg, and it gives the accumulated distance and time en route, leg after leg. If the computer were then taken along on the flight, the pilot could update such parameters as to wind or flight altitude and obtain corrected data for the legs which are affected by these changes.

8
THE NAVTRONIC 1701 AVIATION COMPUTER

The Navtronic 1701 is actually not one but a family of four related computers. There is the basic 1701, the 1701t which includes a timer function, the 1701r which does not have the timer, but does include an area-navigation program, and the 1701tr, which includes both the timer and the area-navigation capability (see Figure 8-1). The unique feature of the 1701 in all its versions is that the operator does not need to remember the different variable values which must be entered into the computer in order to permit it to perform a given computation. The computer is designed to tell the pilot what it needs to know. It does this by activating a tiny red light next to certain printed instructions. Thus, when the light appears next to **SPD**, we know that the computer needs a speed figure in order to be able to go on to the next step.

The Navtronic 1701 consists of a display using LEDs, a 30-key keyboard, two slide switches which are used in conjunction with the weight-and-balance program, and 14 red annunciator lights which represent requests by the computer for additional information. The top 10 keys are the numerical keys from **1** through **0**. The next row includes the function keys: **C/CE** (correct when pressed once, erase when pressed twice), **./:** (decimal point, or to display time in terms of hours:minutes:seconds), **=** and **START** (a dual-function key to be used for equals, and, in the 1701t and tr, to start the timer function), **+/−** and **STOP** (change sign and stop timer key), **ENT** and **T/C** (the enter and timer-function key). Below this row, in the center of the computer, are 5 more dual-function keys: **+** and **ST−NT** (addition and conversion of statute to nautical miles), **−** and **NT−ST** (subtraction and conversion of nautical miles to statute miles), **×** and **°F−°C** (mul-

9

TEXAS INSTRUMENTS' PROGRAMMABLE TI58C and TI59

Much like the Hewlett-Packard HP-41C which is described in Chapter 10, the TI58C and TI59 are portable computer systems with the capability of accepting individual computer programs designed by the operator, in addition to being offered with a preprogrammed aviation module which contains a considerable number of aviation-related programs, some very useful, some not so (see Figure 9-1).

Both models are identical in most respects, but they differ in some. The primary differences are these: The number of memories in which data can be stored is limited to 30 in the TI58 while 60 such memories are available in the TI59. A second difference concerns the ability of the computer to retain information when it has been turned off. Like the HP-41C, the TI58C has a *nonvolatile memory*, meaning that information placed into the computer is retained, even when the switch is turned off. The TI59 has a *volatile memory*, meaning that once it has been turned off, all stored data is lost. Instead the TI59 has the capability of recording data and programs on magnetic cards for later use. These cards are fed into a slot in the computer on the right and are pulled through by a built-in motor. They then appear on the left and the information stored in the computer has been printed electronically on the magnetic surface of the card, where it remains indefinitely unless the card is reused to store a new set of data. When the process is repeated with the same card after the computer has been turned off and then on again, the data is reinserted into the computer and can again be used to run a program.

Both computers have a 45-key keyboard, with most of the keys capable of performing several functions. The top 5 keys, marked **A** through **E** and **A'** through **E'**, are designed to accept specific

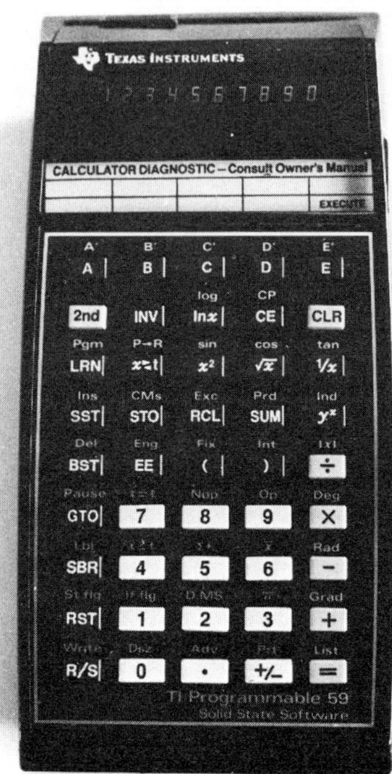

FIGURE 9-1 The Programmable TI59 from Texas Instruments. One of the most sophisticated portable computers on the market.

operator-designated data, quite similar in their function to the **A** through **E** keys on the HP-41C.

As is true of practically all of the systems described in this book, the worst thing about these computers is the instruction book, or rather books. The computer comes with a ¾-inch-thick book entitled *Personal Programming* which, as usual, was written by experts who knew what they were talking about, but personally, and I consider myself to be of at least average intelligence, I found that I read certain portions five or six times, and, since the computer refused to do what the book said it should, I obviously failed to understand some hidden meaning. My guess is that the average pilot, professional or otherwise, would have to be prepared to spend days and weeks practicing in order to be able to make use of even a fraction of the total capabilities of these computers.

The display in both computers uses LED symbols, and all displayed material is in digit form. No letters are used except for a dim *C* which appears throughout the time when the computer is busy "thinking" out a problem. The current is provided by means of rechargeable batteries and the computer will function for about 2 hours between charge cycles. For operations which take longer, or when the card-drive motor is being used repeatedly, it is advisable to keep it plugged into the regular house current through the converter which is provided along with the system.

The total system includes an optional thermal printer, considerably larger than the one associated with the HP-41C, and therefore not exactly portable. But it is a valuable addition to the system, supplying the pilot with a hard copy of the computations performed.

Both computers have a ten-digit display and, like many of the others, they like to come up with a string of endless decimals and decimals of decimals which are fairly useless to the user. By pressing the appropriate keys (**2nd Fix** plus any number from **0** to **9**) the number of decimals can automatically be reduced to anywhere from none to nine. What this means is that a display which would normally look like this: *4.256896741* could be displayed as *4.26* or *4.3* or even just *4*. In this manner, while the actual display is reduced to a group of figures which is easy to comprehend, the actual figure with all its decimals is retained in the computer itself and is used in its entirety in any subsequent computations, even though it is not displayed in that manner.

In addition to the *Personal Programming* book of instructions, purchasers of the special-purpose modules, such as the aviation module (there are twelve such special-purpose modules available from Texas Instruments), also receive a separate book of instructions of how to use the programs stored permanently in the module. In the case of the aviation programs, this book appears to be written in a somewhat more easy-to-comprehend manner, but even it takes a bit of practice.

The insertion of any special-purpose module does not reduce the usefulness of the computer for any other purpose, such as the performance of normal calculator functions, or the design, writing, and execution of any type of personal programs.

How the TI58C/59 functions in performing the normal everyday aviation calculations is described in detail in the chapters appropriate to those functions. In addition, I have used the computer to devise a program designed to determine the fastest and the most economical altitudes for a given flight under certain headwind, tailwind, or no-wind conditions. This is described in detail in the special chapter on the subject of personal programming (see Chapter 33).

The TI59 sells for around $300 without the optional printer, and the TI58 is somewhat less expensive.

Now let's take a quick look, on the following pages, at all twenty-three of the aviation programs which are stored in the aviation module and thus available to the operator at a moment's notice. Let me say right here that, as is true of many of the other computers we're discussing in this book, quite a few of these programs are really of no practical value to the average pilot. Unless we're corporate pilots, stuck with trying to kill hours while sitting at one airport or another, it is rather doubtful that we'd ever make full use of all these programs (except, of course, someone like myself who, as an aviation writer, is frequently faced with the need to figure out what certain types of airplanes can be expected to do under given conditions).

AV-01: AVIATION LIBRARY DIAGNOSTIC

This program simply performs a diagnostic check of the aviation module to see that everything is in working order. It is of no other practical value.

AV-02: FLIGHT PLAN WITH WIND

This program is designed to analyze a planned flight, consisting of one or several legs, even if fuel stops are anticipated. It is a complicated and lengthy program and if the optional printer is available its use will simplify remembering the plethora of data which results. Table 9-1 is a step-by-step breakdown of a four-leg flight, with descriptions of what each step accomplishes and what each displayed value represents (see also Figure 7-2).

TABLE 9-1 PROGRAM AV-02: FLIGHT PLAN WITH WIND

The trip described below is the same one we used in conjunction with the Heathkit OC-1401 and will also use in conjunction with the Hewlett-Packard HP-41C. It goes from Albuquerque to Santa Fe to Las Vegas, New Mexico, to Pueblo, Colorado, and from there to Denver. If the resulting figures for the programs don't work out exactly the same, the primary reason is that the HP-41C makes precise allowances for variations in speed and fuel flow for climb and descent, while in the above program, those differences were based on estimates.

Somewhere in the bowels of the computer this program utilizes the memory-storage positions 00 through 18. In the event that the optional printer is not used, the operator may wish to store such subtotals as the time and fuel for each leg into separate memories (numbers 19 through 59) by pressing STO 59 (or whatever memory position is available) in order to be able to recall this data at a later time. This can be accomplished without compromising the program.

Data to be entered	Press key(s)	Display	Description of the display meaning
Identify program	**2nd**	*0*	
	Pgm	*0*	
	02	*0*	
Initialize program	**SBR**	*0*	
	CLR	*0.0000*	
Time of planned takeoff: 12:00:00	**12.0**	*12.0*	Estimated time of departure
Move to key position	**2nd**	*12.0*	
	E'	*12.0000*	Appears briefly, then changes
		0.0000	to this which represents total time until now
	R/S	*0.0000*	Total fuel until now
The next entries represent data for the first leg of the trip:			
Magnetic variation	**12**	*12*	
	+/−	*−12*	The easterly variation
Move to key position	**2nd**	*−12*	
	A'	*−12.0000*	
Wind direction	**310**	*310*	Wind direction (true)

(Continued)

TABLE 9-1 (*Continued*)

Data to be entered	Press key(s)	Display	Description of the display meaning
Move to key position	**2nd**	310	
	B'	310.0000	
Wind velocity	**45**	45	Wind velocity, kn
Move to key position	**2nd**	45	
	C'	45.0000	
True airspeed	**150**	150	TAS, kn
Move to key position	**2nd**	150	
	D'	150.0000	
Fuel flow (average for climb and cruise)	**24**	24	Fuel flow, gal/h
	R/S	24.0000	
True course	**43**	43	True course 043°
Move to key position	**A**	43.0000	
Distance	**47**	47	Distance, nmi
Move to key position	**B**	47.0000	
Now determine true heading	**C**	43.0000	Appears briefly, then changes
		25.5671	to the true heading to be flown to achieve the true course
	R/S	13.5671	The magnetic heading
	D	47.0000	The distance appears briefly, then changes to
		145.4654	which represents the ground speed
	R/S	0.1923	Time spent to reach the end of leg 1
	E	7.7544	Fuel used so far
	R/S	12.1923	Estimated time of arrival at the end of leg 1
	2nd		
	E'	12.1923	Appears briefly, then changes
		0.1923	to this, again representing the total ETE
	R/S	7.7544	Repeats total fuel so far
The next entries represent data for the second leg of the trip:			
Magnetic variation	**12**	12	
	+/−	−12	
Move to key position	**2nd**	−12	
	A'	−12.0000	
Wind direction at higher altitude	**315**	315	Wind direction (true)
Move to key position	**2nd**	315	
	B'	315.0000	
Wind velocity	**55**	55	Wind velocity, kn
Move to key position	**2nd**	55	
	C'	55.0000	

TABLE 9-1 (*Continued*)

Data to be entered	Press key(s)	Display	Description of the display meaning
True airspeed	**150**	*150*	TAS, kn
Move to key position	**2nd**	*150*	
	D'	*150.0000*	
Fuel flow in cruise	**22**	*22*	Fuel flow, gal/h
	R/S	*22.0000*	
True course	**68**	*68*	True course 068°
Move to key position	**A**	*68.0000*	
Distance	**46**	*46*	Distance, nmi
Move to key position	**B**	*46.0000*	
	C	*68.0000*	Appears briefly, then changes
		48.2742	to the true heading to be flown to achieve the true course
	R/S	*36.2742*	The magnetic heading
	D	*46.0000*	Appears briefly, then changes
		162.6880	to the ground speed, kn
	R/S	*0.1658*	ETE for leg 2
	E	*6.2205*	Fuel used for leg 2
	R/S	*12.3621*	ETA at end of leg 2

If we wanted to land there and stay a while, we could add to this the time of the layover in hours and decimals by pressing: **+ time =**. Without a layover we go on with

Data to be entered	Press key(s)	Display	Description of the display meaning
	2nd	*12.3621*	
	E'	*0.3621*	Total time for leg 1 and 2
	R/S	*13.9749*	Total fuel for leg 1 and 2

The next entries represent data for the third leg of the trip:

Data to be entered	Press key(s)	Display	Description of the display meaning
Magnetic variation	**12**	*12*	
	+/−	*−12*	Easterly variation
Move to key position	**2nd**	*−12*	
	A'	*−12.0000*	
Wind direction	**315**	*315*	Wind direction (true)
Move to key position	**2nd**	*315*	
	B'	*315.0000*	
Wind velocity	**55**	*55*	Wind velocity, kn
Move to key position	**2nd**	*55*	
	B'	*55.0000*	
True airspeed	**150**	*150*	TAS, kn
Move to key position	**2nd**	*150*	
	D'	*150.0000*	
Fuel flow in cruise	**22**	*22*	Fuel flow, gal/h
	R/S	*22.0000*	
True course	**358**	*358*	True course

(Continued)

Data to be entered	Press key(s)	Display	Description of the display meaning
Move to key position	A	358.0000	
Distance	162	162	Distance, nmi
Move to key position	B	162.0000	
	C	358.0000	Appears briefly, then changes
		343.5186	to the true heading
	R/S	331.5186	The magnetic heading
	D	162.0000	Appears briefly, then changes
		105.0099	to the ground speed
	R/S	1.3234	ETE for leg 3
	E	33.9397	Fuel used for leg 3
	R/S	14.0855	ETA at end of leg 3 (24-h clock)
	2nd		
	E'	14.0855	Appears briefly, then changes
		2.0855	to the total time so far
	R/S	47.9146	Total fuel burned so far
The next entries represent data for the fourth leg of the trip:			
Magnetic variation	12	12	
	+/−	−12	Easterly variation
Move to key position	2nd	−12	
	A'	−12.0000	
Wind direction	315	315	Wind direction (true)
Move to key position	2nd	315	
	B'	315.0000	
Wind velocity	55	55	Wind velocity, kn
Move to key position	2nd	55	
	C'	55.0000	
True airspeed	150	150	TAS, kn
Move to key position	2nd	150	
	D'	150.0000	
Fuel flow average for cruise and descent	18	18	Fuel flow average gal/h
	R/S	18.0000	
True course	338	338	True course
Move to key position	A	338.0000	
Distance	95	95	Distance for leg 4, nmi
Move to key position	B	95.0000	
	C	338.0000	Appears briefly, then changes
		329.7630	to the true heading
	R/S	317.7630	The magnetic heading
	D	95.0000	Appears briefly, then changes
		97.8248	to the ground speed, kn
	R/S	0.5816	ETE
	E	17.4802	Fuel burned
	R/S	15.0711	ETA (24-h clock)
	2nd	15.0711	
	E'	15.0711	Appears briefly, then changes
		3.0711	to total time for trip
	R/S	65.3948	Total fuel used for trip

AV-03: FLIGHT PLAN AND VERIFICATION

This is very similar to AV-02 except that, instead of computing the ground speed resulting from a given airspeed, it computes the TAS needed in order to achieve the desired ground speed under given wind conditions at different altitudes. Table 9-2 shows how this program works in practice.

TABLE 9-2   PROGRAM AV-03: FLIGHT PLAN AND VERIFICATION

Data to be entered	Press key(s)	Display	Description of the display meaning
Identify program	2nd	0	
	Pgm	0	
	03	0	
Initialize program	SBR	0	
	CLR	0.0000	
Enter takeoff time	12.0	12.0	Takeoff at noon
Move to key position	A	12.0000	
Enter distance	245	245	Distance to be flown
Move to key position	D	0.0000	
Enter arrival time	13.4	13.4	Time we want to arrive
Move to key position	2nd	13.4	
	E'	13.4000	
Enter wind direction	345	345	Wind from 345°
Move to key position	2nd	345	
	B'	345.0000	
Enter wind speed	34	34	Wind velocity, kn
Move to key position	2nd	34	
	C'	34.0000	
Enter true course	285	285	Course 285° (true)
Move to key position	2nd	285	
	D'	295.1785	In order to make good the 285° course, we have to steer a true heading of 295.1785°
	R/S	166.6223	The TAS in knots which must be achieved in order to arrive at the desired time

AV-04: LONG-RANGE FLIGHT PLAN

This program is so incredibly complicated that it requires the use of the optional printer in order to be of use. Table 9-3 is a diagnosis of this program, showing each keystroke, each display, and the data which will appear on the hard copy produced by the optional printer.

TABLE 9-3   PROGRAM AV-04: LONG-RANGE FLIGHT PLAN

Using this program requires the use of the optional printer, because the desired results are not displayed in the computer display, but only printed out. This program can accommodate up to 44 waypoints when the TI59 is used, or 24 waypoints when the TI58C is used.

Data to be entered	Press key(s)	Display	Printout		Description of the display meaning
Identify program	**2nd**	*0*			
	Pgm	*0*			
	04	*0*			
Initialize program	**SBR**	*0*			
	CLR	*0.0000*	LONG RANGE FLT PLAN		
			0.0000	WP	Waypoint 0
Latitude, WP 0	**49.2036**	*49.2036*			
Move to key position	**A**	*1.000000*	49.2036	LAT	
Longitude, WP 0	**123.1024**	*123.1024*			
Move to key position	**A**	*1.000000*	123.1024	LON	
			1.0000	WP	Waypoint 1
Latitude, WP 1	**48.12**	*48.12*			
Move to key position	**A**	*1.000000*	48.1200	LAT	
Longitude, WP 1	**120.46**	*120.46*			
Move to key position	**A**	*1.000000*	120.4600	LON	
			2.0000	WP	Waypoint 2
Latitude, WP 2	**47.31**	*47.31*			
Move to key position	**A**	*1.000000*	47.3100	LAT	
Longitude, WP 2	**114.55**	*114.55*			
Move to key position	**A**	*1.000000*	114.5500	LON	
			3.0000	WP	Waypoint 3
Latitude, WP 3	**45.06**	*45.06*			
Move to key position	**A**	*1.000000*	45.0600	LAT	
Longitude, WP 3	**110.2306**	*110.2306*			
Move to key position	**A**	*1.000000*	110.2306	LON	
			4.0000	WP	Waypoint 4
Latitude, WP 4	**43.1612**	*43.1612*			
Move to key position	**A**	*1.000000*	43.1612	LAT	
Longitude, WP 4	**104.18**	*104.18*			
Move to key position	**A**	*1.000000*	104.1800	LON	
			5.0000	WP	Waypoint 5

TABLE 9-3 (*Continued*)

Data to be entered	Press key(s)	Display	Printout		Description of the display meaning
Latitude, WP 5	**40.28**	*40.28*			
Move to key position	**A**	*1.000000*	40.2800	LAT	
Longitude, WP 5	**101.1654**	*101.1654*			
Move to key position	**A**	*1.000000*	101.1654	LON	
			6.0000	WP	Waypoint 6
Latitude, WP 6	**38.0718**	*38.0718*			
Move to key position	**A**	*1.000000*	38.0718	LAT	
Longitude, WP 6	**96.4754**	*96.4754*			
Move to key position	**A**	*1.000000*	96.4754	LON	
Enter ground speed	**230**	*230*			
Move to key position	**B**	*230.0000*	230.0000	GS	Ground speed
Enter fuel on board	**100**	*100*			
Move to key position	**C**	*100.0000*	100.0000	FUEL	100 gal on board
Enter fuel flow	**15**	*15*			
Move to key position	**D**	*15.0000*	15.0000	BURN	Fuel flow, gal/h
Enter ETD	**10.45**	*10.45*			
Move to key position	**E**	*2.0000*	10.4500	ETD	Takeoff time: 10:45
			1316.8427	DIST	Total distance, nmi
			5.4331	ETE	Time en route
			16.2831	ETA	Arrival time estimated as 16:28:31
			85.8810	EFR	Total amount of fuel burned
			14.1190	EFL	Amount of fuel remaining

AV-05: ATMOSPHERE, SPEED, TEMPERATURE, AND ALTITUDE

This program, also described in part in some of the chapters on basic aviation problems, is designed to compute TAS, Mach number, density altitude and true air temperature. Table 9-4 shows and explains each program step and the different results.

TABLE 9-4   PROGRAM AV-05: ATMOSPHERE, SPEED, TEMPERATURE, AND ALTITUDE

Data to be entered	Press key(s)	Display	Description of the display meaning
Identify program	**2nd**	*0*	
	Pgm	*0*	
	05	*0*	
Initialize program	**SBR**	*0*	
	CLR	*0.00*	
Flight altitude	**41000**	*41000*	41,000 feet (FL 410)
Move to key position	**A**	*41000.00*	Pressure altitude
	2nd	*41000.00*	
	B'	*−66.22*	True temperature, °C
	2nd	*−66.22*	
	C'	*0.85*	Relative speed of sound
	×	*0.85*	
Speed of sound	**661.5**	*661.5*	
	=	*559.55*	TAS, kn
	2nd	*559.55*	
	D'	*0.18*	Relative atmospheric pressure
	×	*0.18*	
Standard atmospheric pressure	**29.92**	*29.92*	
	=	*5.25*	Actual atmospheric pressure, in Hg
	2nd	*5.25*	
	E'	*0.24*	Relative atmospheric density

AV-06: FREEZE LEVEL; LOWEST FLIGHT LEVEL

This program predicts freezing levels and the lowest usable flight level based on known temperatures and the prevailing barometric pressure. Table 9-5 explains how it works.

TABLE 9-5   PROGRAM AV-06: PREDICTING FREEZING LEVELS AND LOWEST USABLE FLIGHT LEVELS

Data to be entered	Press key(s)	Display	Description of the display meaning
Identify program	**2nd**	0	
	Pgm	0.	
	06	0.	
Initialize program	**SBR**	0.	
	CLR	0.	
Enter temperature °F	**45**	45	The computer changes the
Move to key position	**2nd**	45	temperature reading to
	A'	7.22	°C; if °C are entered originally, press **A** instead of **2nd A'**
Enter altitude	**7000**	7000	7000 ft
Move to key position	**B**	7000.	
	C	10611	The freezing level in clear and dry air is 10,611 ft
	D	11815	The freezing level in clouds is 11,815 ft
Enter altimeter setting	**30.03**	30.03	30.03 inHg
	E	18000	Lowest usable flight level for high-altitude IFR aircraft is 18,000 ft

AV-07: WIND COMPONENTS AND AVERAGE VECTOR

This program consists of two parts. When given wind velocity, wind direction, magnetic heading, and the magnetic variation, it calculates wind components and vectors. Given information relative to wind vectors en route, it will calculate the average vector for the entire flight. Table 9-6 explains how.

TABLE 9-6   PROGRAM AV-07: WIND COMPONENTS AND AVERAGE VECTOR

Data to be entered	Press key(s)	Display	Description of the display meaning
Identify program	**2nd**	0	
	Pgm	0	
	07	0	
Initialize program	**SBR**	0	
	CLR	0.00	
Magnetic variation	**12**	12	
	+/−	−12	Easterly variation
Move to key position	**2nd**	−12	
	A'	−12.00	
Wind direction	**345**	345	Wind direction (true)
Move to key position	**2nd**	345	
	B'	345.00	
Wind velocity	**34**	34	Wind speed is 34 kn
Move to key position	**2nd**	34	
	C'	34.00	
Magnetic heading	**296**	296	We're steering 296°
Move to key position	**2nd**	296	
	D'	296.00	
	A	−27.15	Head wind component, kn
	B	−20.46	Right crosswind component, kn

AVERAGE VECTOR

Wind direction, first part of flight	**345**	345	
Move to key position	**2nd**	345	
	B'	345.00	
Wind velocity, first part of flight	**34**	34	
Move to key position	**2nd**	34	
	C'	34.00	
Distance of first part of flight	**85**	85	
Move to key position	**2nd**	85	
	E'	85.00	

TABLE 9-6 (*Continued*)

Data to be entered	Press key(s)	Display	Description of the display meaning
Wind direction, second part of flight	245	245	
Move to key position	2nd	245	
	B'	245.00	
Wind velocity, second part of flight	15	15	
Move to key position	2nd	15	
	C'	15.00	
Distance of second part of flight	46	46	
Move to key position	2nd	46	
	E'	46.00	
Wind direction, third part of flight	215	215	
Move to key position	2nd	215	
	B'	215.00	
Wind velocity, third part of flight	10	10	
Move to key position	2nd	10	
	C'	10.00	
Distance of third part of flight	70	70	
Move to key position	2nd	70	
	E'	70.00	
	D	317.35	Average wind direction
	E	13.03	Average wind velocity
	2nd	13.03	
	C'	13.03	
	D	317.35	
	2nd	317.35	
	B'	317.35	
Magnetic variation	12	12	
	+/−	−12	
Move to key position	2nd	−12	
	A'	−12.00	
Magnetic heading	296	296	
Move to key position	2nd	296	
	D'	296.00	
	A	−12.86	Head wind, kn
	B	−2.12	Right crosswind, kn

AV-08: WIND TRIANGLE

This program can be used to calculate wind direction and velocity or magnetic course and ground speed, based on known magnetic variation, magnetic heading, and TAS. Table 9-7 explains.

TABLE 9-7   PROGRAM AV-08: WIND TRIANGLE

Data to be entered	Press key(s)	Display	Description of the display meaning
Identify program	**2nd**	*0*	
	Pgm	*0*	
	08	*0*	
Initialize program	**SBR**	*0*	
	CLR	*0.00*	
Magnetic variation	**12**	*12*	Easterly variation
	+/−	*−12*	
Move to key position	**2nd**	*−12*	
	A'	*−12.00*	
Magnetic heading	**296**	*296*	
Move to key position	**2nd**	*296*	
	D'	*296.00*	
True airspeed	**150**	*150*	
Move to key position	**2nd**	*150*	
	E'	*150.00*	

To calculate wind direction and velocity enter:

Data to be entered	Press key(s)	Display	Description of the display meaning
Magnetic course	**285**	*285*	
Move to key position	**B**	*285.00*	
Ground speed	**130**	*130*	
Move to key position	**E**	*130.00*	
	A	*−1.00*	
	2nd	*−1.00*	
	B'	*355.93*	Wind direction (true)
	A	*−1.00*	
	2nd	*−1.00*	
	C'	*33.41*	Wind velocity, kn

TABLE 9-7 (*Continued*)

Data to be entered	Press key(s)	Display	Description of the display meaning

To calculate magnetic course and ground speed, start after the TAS was entered and enter:

Data to be entered	Press key(s)	Display	Description of the display meaning
Wind direction	355	355	
Move to key position	2nd	355	
	B'	355.00	
Wind velocity	33	33	
Move to key position	2nd	33	
	C'	33.00	
	A	−1.00	
	B	285.28	Magnetic course
	A	−1.00	
	C	129.76	Ground speed, kn

AV-09: DEAD RECKONING

When known latitude, longitude, and takeoff time as well as true course and ground speed are provided to the computer, this program will determine the position of the aircraft in terms of latitude and longitude at a later time. Table 9-8 shows how.

TABLE 9-8 PROGRAM AV-09: DEAD RECKONING

Data to be entered	Press key(s)	Display	Description of the display meaning
Identify program	2nd	0	
	Pgm	0	
	09	0	
Initialize program	SBR	0	
	CLR	0.0000	
Takeoff time h:min:s	12.0	12.0	12:00:00 (noon)
Move to key position	2nd	12.0	
	A'	12.0000	
Starting latitude	41.16	41.16	
Move to key position	B	41.2667	The computer has changed degrees, minutes, and seconds to degrees and decimals
Starting longitude	87.48	87.48	
Move to key position	2nd	87.48	
	B'	87.8000	The same took place with the longitude
Enter true course	285	285	
Move to key position	C	285.0000	
Ground speed	130	130	
Move to key position	D	130.0000	
Current time h:min:s	14.3230	14.3230	2:32:30 p.m.
Move to key position	2nd	14.3230	
	D'	14.5417	Computer has changed time to decimal format
	E	42.4131	Present position latitude is 43° 41' 31"
	2nd	42.4131	
	E'	94.5722	Present position longitude is 94° 57' 22"

AV-10: RHUMBLINE NAVIGATION

This program is described in detail in Chapter 28 "Area Navigation."

AV-11: GREAT CIRCLE FLYING

This program, too, is examined closely in Chapter 28.

AV-12: LINE-OF-SIGHT DISTANCE AND ALTITUDE DME SPEED CORRECTION

Even though this program is explained in some detail in Chapter 31 "DME Slant-Range Correction," Table 9-9 shows all of its capabilities which include determining the reception distance for very high-frequency (VHF) transmitters which is available at varying altitudes.

TABLE 9-9   PROGRAM LINE-OF-SIGHT DISTANCE AND ALTITUDE DME SPEED CORRECTION

DME slant-range speed corrections can be made with this program and are explained in Chapter 31.

Data to be entered	Press key(s)	Display	Description of the display meaning
Identify program	**2nd**	*0*	
	Pgm	*0*	
	12	*0*	
Initialize program	**SBR**	*0*	
	CLR	*0.*	
Terrain altitude	**2825**	*2825*	Highest terrain altitude between the aircraft and the station
Move to key position	**A**	*2825.*	
Enter transmitter altitude	**8**	*8*	Transmitter is 8 ft msl
Move to key position	**B**	*8.*	

To calculate line of sight for a given aircraft altitude enter:

Data to be entered	Press key(s)	Display	Description of the display meaning
Aircraft altitude:	**5500**	*5500*	
Move to key position	**C**	*5500.*	
	E	*5500.*	
	D	*112.*	Line-of-sight transmission distance is 112 nmi

To calculate required altitude to permit transmission over a given distance, enter:

Data to be entered	Press key(s)	Display	Description of the display meaning
Distance	**112**	*112*	
Move to key position	**D**	*112.*	
	E	*112.*	
	C	*5547*	The required altitude, msl

AV-13: POSITION AND NAVIGATION BY ONE VOR

This is a program, the usefulness of which escapes me. But I may be quite wrong, so Table 9-10 shows what it does, how and why.

TABLE 9-10   PROGRAM AV-13: POSITION AND NAVIGATION BY ONE YEAR

Data to be entered	Press key(s)	Display	Description of the display meaning
Identify program	**2nd**	*0*	
	Pgm	*0*	
	13	*0*	
Initialize program	**SBR**	*0*	
	CLR	*0.0000*	
Magnetic variation	**12**	*12*	
	+/−	*−12*	Easterly variation
Move to key position	**2nd**	*−12*	
	A'	*−12.0000*	
Wind direction	**345**	*345*	
Move to key position	**2nd**	*345*	
	B'	*0.0000*	
Wind velocity	**34**	*34*	
Move to key position	**2nd**	*34*	
	B'	*0.0000*	
VOR radial to destination or waypoint	**245**	*245*	
Move to key position	**2nd**	*245*	
	C'	*0.0000*	
Distance from VOR to destination or WP	**87**	*87*	
Move to key position	**2nd**	*87*	
	C'	*0.0000*	
Magnetic heading	**285**	*285*	
Move to key position	**2nd**	*285*	
	D'	*285.0000*	
True airspeed	**150**	*150*	
Move to key position	**2nd**	*150*	
	E'	*150.0000*	
Time when radial from the VOR was first read, hr:min:s	**12.3030**	*12.3030*	
Move to key position	**A**	*12.5083*	Time in decimal format
Radial of the VOR on which the aircraft is positioned at that first time reading	**90**	*90*	

TABLE 9-10 (*Continued*)

Data to be entered	Press key(s)	Display	Description of the display meaning
Move to key position	**B**	*90.0000*	
Time when radial from VOR was read the second time, h:min:s	**12.4224**	*12.4224*	
Move to key position	**A**	*12.7067*	Time in decimal format
Radial of the VOR on which the aircraft is positioned at that second time reading	**98**	*98*	
Move to key position	**B**	*98.0000*	
	C	*12.1540*	Distance between the VOR and the aircraft at the second time reading
	D	*357.2068*	Magnetic course to the destination or waypoint
	R/S	*246.9849*	Distance to the destination or waypoint, nmi
	R/S	*129.7338*	Ground speed, kn
	R/S	*14.3637*	Estimated time at which the aircraft will arrive at the destination or waypoint (2:36:37 p.m.)

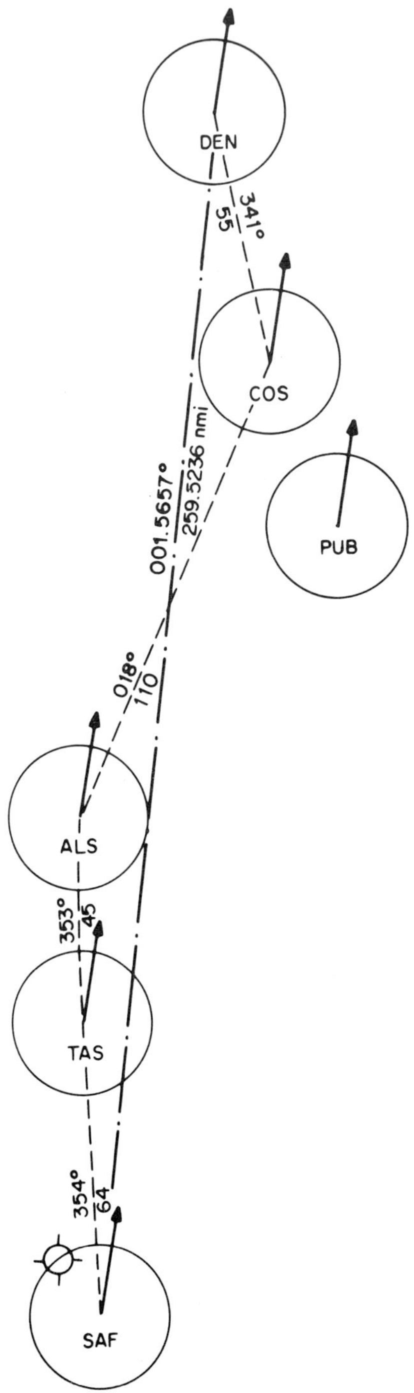

FIGURE 9-2 These are the VORs which have been used in running the AREA NAVIGATION program.

AV-14, AV-15, and AV-16: AREA NAVIGATION

This is an outrageously complicated three-part program which allows for the storage of waypoints and the determination of all navigational data associated with flying a given flight by means of area navigation. While it can be used in flight, my own feeling is that any degree of turbulence would make it pretty nearly impossible to execute it without making some mistakes somewhere along the way. Figure 9-2 shows and Table 9-11 describes what is involved in running this program.

TABLE 9-11   PROGRAMS AV-14, AV-15, AND AV-16: AREA NAVIGATION

This program consists of three parts, one of which, AV-15 must be executed before using either AV-14 or AV-16. AV-15, the AREA NAV LOAD MODULE defines the relationship between a group of VORs or other navigation aids or points of known geographic location, which represent the area in which a flight is to take place. This information is then used by the other two programs. (See Figure 9-2.)

Data to be entered	Press key(s)	Display	Description of the display meaning
AV-15:			
Identify program	**2nd**	*0*	
	Pgm	*0*	
	15	*0*	
Initialize program	**SBR**	*0*	
	CLR	*0.*	
VOR 1	**1**	*1*	TAS (Taos, New Mexico)
Place in key position	**A**	*1*	
VOR 0	**0**	*0*	SAF (Sante Fe, New Mexico)
Place in key position	**B**	*0*	
Radial 0–1	**354**	*354*	The radial from SAF to TAS (see Figure 9-2)
Place in key position	**C**	*354*	
Distance 0–1	**64**	*64*	The distance from SAF to TAS (see Figure 9-2)
Place in key position	**D**	*64*	
VOR 2	**2**	*2*	ALS (Alamosa, Colorado)
Place in key position	**A**	*2*	
VOR 1	**1**	*1*	TAS (Taos, New Mexico)
Place in key position	**B**	*1*	
Radial 1–2	**353**	*353*	Radial from TAS to ALS
Place in key position	**C**	*353*	
Distance 1–2	**45**	*45*	Distance from TAS to ALS
Place in key position	**D**	*45*	

(Continued)

TABLE 9-11 (*Continued*)

Data to be entered	Press key(s)	Display	Description of the display meaning
VOR 3	**3**	*3*	COS (Colorado Springs, Colorado)
Place in key position	**A**	*3*	
VOR 2	**2**	*2*	ALS (Alamosa, Colorado)
Place in key position	**B**	*2*	
Radial 2–3	**18**	*18*	018° ALS to COS
Place in key position	**C**	*18*	
Distance 2–3	**110**	*110*	110 nmi ALS to COS
Place in key position	**D**	*110*	
VOR 4	**4**	*4*	DEN (Denver, Colorado)
Place in key position	**A**	*4*	
VOR 3	**3**	*3*	COS (Colorado Springs, Colorado)
Place in key position	**B**	*3*	
Radial 3–4	**341**	*341*	341° COS to DEN
Place in key position	**C**	*341*	
Distance 3–4	**55**	*55*	55 nmi COS to DEN
Place in key position	**D**	*55*	

At this point the relationship between the five VORs has been entered into the computer. If it is to be used at a later time, this information can be recorded on one of the magnetic cards (with the TI59 only) by this series of actions:

CLR	*0*	
4	*4*	
2nd	*4*	
Write		The card is now fed into the lower slot and the information is recorded on it

Now, in order to use this information for a straight-line (RNAV) flight from Santa Fe to Denver, we continue with program AV-16. If the computer has been turned off in the interim, we first reenter the above information into it by pressing CLR and then feeding the card into the lower slot. The display will show 4 to indicate that the card has been satisfactorily read.

AV-16:

Identify program	**2nd**	*0*
	Pgm	*0*
	16	*0*
Initialize program	**SBR**	*0*
	CLR	*0.0000*
Expected ground speed	**150**	*150*
Place in key position	**2nd**	*150*
	D'	*150.0000*

TABLE 9-11 (*Continued*)

Data to be entered	Press key(s)	Display	Description of the display meaning
Takeoff time,			
h:min:s	**12.30**	*12.30*	
Place in key position	**A**	*12.5000*	Time in decimal format
VOR 0	**0**	*0*	
Place in key position	**2nd**	*0*	
	A'	*0.0000*	
Position from 0	**330**	*330*	We're sitting on the airport in Santa Fe which is on the 330° radial some 5 nmi from the Santa Fe VOR
Place in key position	**2nd**	*330*	
	B'	*0.0000*	
VOR 2	**2**	*2*	
Place in key position	**2nd**	*2*	
	A'	*0.0000*	
Position from 2	**355**	*355*	Still on the airport, we're on the 355° bearing to the Alamosa VOR
Place in key position	**2nd**	*355*	
	C'	*0.0000*	
VOR 4 (destination)	**4**	*4*	
	B	*259.5236*	The straight-line distance from the Santa Fe airport to the Denver VOR, nmi
	D	*1.5657*	001.5657° is the magnetic course
	C	*14.1349*	The expected arrival time is 2:13:49 p.m.

Program AV-14 is identical in character, but it is designed for aircraft equipped with DME, and, as a result, uses less keystrokes.

AV-14:

Identify program	**2nd**	*0*	
	Pgm	*0*	
	14	*0*	
Initialize program	**SBR**	*0*	
	CLR	*0.0000*	
Expected ground speed	**150**	*150*	
Place in key position	**2nd**	*150*	
	D'	*150.0000*	
Takeoff time (h:min:s)	**12.30**	*12.30*	
Place in key position	**A**	*12.5000*	Time in decimal format
VORTAC 0	**0**	*0*	SAF (Santa Fe, New Mexico)

(*Continued*)

TABLE 9-11 (*Continued*)

Data to be entered	Press key(s)	Display	Description of the display meaning
Place in key position	**2nd**	*0*	
	A'	*0.0000*	
Radial from VORTAC	**330**	*330*	Airport from SAF VORTAC
Place in key position	**2nd**	*330*	
	B'	*330.0000*	
DME distance	**6**	*6*	Actual distance
Place in key position	**2nd**	*6*	
	C'	*6.0000*	
VORTAC 4	**4**	*4*	DEN (Denver, Colorado)
	B	*259.8295*	Distance to destination in nmi
	D	*1.5234*	001.5234° is the magnetic course
	C	*14.1356*	Estimated time of arrival is 2:13:56 p.m.

AV-17: COURSE CORRECTION

This program is described in the instructions as being designed to calculate the corrected magnetic course and the distance to go for an aircraft that has strayed a known distance off course. Frankly, if I know how far I am to the right or left of course, I can't imagine why I would need a computer to get back on course. Still, Table 9-12 shows how this works.

TABLE 9-12   PROGRAM AV-17: COURSE CORRECTION

Data to be entered	Press key(s)	Display	Description of the display meaning
Identify program	2nd	0	
	Pgm	0	
	17	0	
Initialize program	SBR	0	
	CLR	0.00	
Distance off course	15	15	
	+/−	−15	15 nmi to the right off course (+ is left, − is right)
Place in key position	A	−15.00	
Distance, total flight	245	245	245 nmi from departure to destination point
Place in key position	B	245.00	
Distance flown	125	125	Estimated distance flown so far
Place in key position	2nd	125	
	C'	124.10	Distance from start to intended checkpoint
Intended course	285	285	Intended course was 285° magnetic
Place in key position	D'	285.00	
	E	277.93	The corrected magnetic course to arrive at the intended destination
	2nd	277.93	
	E'	121.83	Distance to go, nmi

AV-18: RATE OF CLIMB; TURN PERFORMANCE

The rate of climb (ROC) portion of this program is explained in detail in Chapter 27. But since this program also deals with turn performance, Table 9-13 explains the computation of stall speed, diameter of a turn, and the G-force at a given bank angle and TAS.

TABLE 9-13 PROGRAM AV-18: RATE OF CLIMB; TURN PERFORMANCE

The rate-of-climb portion of this program has been explained in detail in Chapter 27. Therefore we'll examine here only the portion dealing with turn performance.

Data to be entered	Press key(s)	Display	Description of the display meaning
Identify program	2nd	0	
	Pgm	0	
	18	0	
Initialize program	SBR	0	
	CLR	0.0000	
True airspeed	150	150	
Move to key position	2nd	150	
	B'	0.0000	
Degree of bank	15	15	
Move to key position	2nd	15	
	C'	0.0000	
Normal stall speed	60	60	
Move to key position	2nd	60	
	D'	0.0000	
	2nd	0.0000	
	E'	1.0353	G-force at 15° bank at 150 kn
	C	2.4547	Diameter of a 360° circle flown at a 15° bank at a TAS of 150 kn
	E	61.0491	Stall speed in knots at a 15° bank
	D	0.0304	It takes 3 min and 4 s to complete a 360° turn

AV-19: GENERAL WEIGHT AND BALANCE

Even though weight-and-balance calculations are a matter of simple arithmetic and can be performed by the calculator portion of the computer with ease, a special program has been designed for the purpose. Table 9-14 shows how that works.

TABLE 9-14   PROGRAM AV-19: GENERAL WEIGHT AND BALANCE

Data to be entered	Press key(s)	Display	Description of the display meaning
Identify program	**2nd**	0	
	Pgm	0	
	19	0	
Initialize program	**SBR**	0	
	CLR	0.00	
Empty weight	**2000**	2000	2000 lb
Move to key position	**A**	2000.00	
Moment	**68634.3**	68634.3	
Move to key position	**2nd**	68634.3	
	B'	68634.30	
Fuel, gal	**80**	80	
Move to key position	**2nd**	80	
	C'	480.00	Fuel, lb
	A	480.00	
Moment	**48**	48	
Move to key position	**B**	23040.00	
Weight of oil	**15**	15	
Move to key position	**A**	15.00	
Moment	**300**	300	
	+/−	−300	Moment −300
Move to key position	**2nd**	−300	
	B'	−300.00	
Pilot and copilot	**340**	340	
Move to key position	**A**	340.00	
Moment	**36**	36	
Move to key position	**B**	12240.00	
Rear-seat passengers	**340**	340	
Move to key position	**A**	340.00	
Moment	**70**	70	
Move to key position	**B**	115600.00	
Baggage weight	**100**	100	
Move to key position	**A**	100.00	
Moment	**95**	95	
Move to key position	**B**	10000.00	
	C	3275.00	Total aircraft weight
	D	229214.30	Total moment
	E	69.99	Center of gravity

AV-20: CUSTOMIZED WEIGHT AND BALANCE

This program serves the same purpose as the one above, except that it requires that all known weight-and-balance parameters for a given aircraft are permanently stored on one of the magnetic cards. Table 9-15 explains.

TABLE 9-15 PROGRAM AV-20: CUSTOMIZED WEIGHT AND BALANCE

The idea of this program is that if the same aircraft is always flown, it will save time and effort to permanently store the empty-weight and moment figures in the form of magnetic cards (TI59 only).

Data to be entered	Press key(s)	Display	Description of the display meaning
Initialize	**CLR**	0	
	2nd	0	
	CMs	0.00	
Moment, fuel	**48**	48	in
Store in memory	**STO**	48.00	
	0	48.00	
	1	48.00	
Baggage, moment	**95**	95	in
Store in memory	**STO**	95.00	
	0	95.00	
	3	95.00	
Moment, front seats	**36**	36	in
Store in memory	**STO**	36.00	
	0	36.00	
	5	36.00	
Moment, 2d row seats	**70**	70	in
	STO	70.00	
	0	70.00	
	6	70.00	
Empty moment	**68634.3**	68634.3	in · lb
Store in memory	**STO**	68634.30	
	0	68634.30	
	8	68634.30	
Empty weight	**2000**	2000	
Store in memory	**STO**	2000.00	
	0	2000.00	
	9	2000.00	

To record this information on a magnetic card, press INV 2nd Fix 4 2nd Write. Then feed in the card and the display will come up with 4 to show that it has been printed.

TABLE 9-15   (*Continued*)

Data to be entered	Press key(s)	Display	Description of the display meaning

Now, any time a weight-and-balance computation for that particular aircraft has to be performed, simply press CLR and feed in the card. The display will show 4 to show that the information has been transferred. After that we are ready to use the actual program AV-20:

Data to be entered	Press key(s)	Display	Description of the display meaning
AV-20:			
Identify program	**2nd**	0	
	Pgm	0	
	20	0	
Initialize program	**SBR**	0	
	CLR	0.00	
Fuel, gal	**80**	80	
Move to key position	**A**	80.00	
Baggage weight	**100**	100	
Move to key position	**B**	100.00	
Pilot, copilot, weight	**340**	340	
Move to key position	**2nd**	340	
	C'	340.00	
Rear passengers, weight	**340**	340	
Move to key position	**2nd**	340	
	D'	340.00	
	C	3260.00	Total weight of the aircraft
	D	119074.30	Total moment, in · lb
	E	36.53	Center of gravity

AV-21: PILOT UNIT CONVERSIONS

This program is designed to perform the conversions (other than time) which pilots are frequently called upon to perform. Table 9-16 shows how that works.

TABLE 9-16 PROGRAM AV-21: PILOT UNIT CONVERSIONS

All parameters to be converted are listed on the AV-21 card which is placed into the upper slot. Conversions are accomplished by simply pressing the appropriate keys.

Data to be entered	Press key(s)	Display	Description of the display meaning
Identify program	**2nd**	*0*	
	Pgm	*0*	
	21	*0*	
Initialize program	**SBR**	*0*	
	CLR	*0.*	
Value to be converted	**10**	*10*	
	2nd	*10*	
	A'	*−12.22*	10°F = −12.22°C
Unit to convert	**10**	*10*	
	2nd	*10*	
	E'	*10.00*	
	2nd	*10.00*	
	A'	*50.00*	10°C = 50°F
Unit to convert	**10**	*10*	
	2nd	*10*	
	B'	*60.00*	10 gal = 60 lb
Unit to convert	**10**	*10*	
	2nd	*10*	
	E'	*10.00*	
	2nd	*10.00*	
	B'	*1.67*	10 lb = 1.67 gal
Unit to convert	**10**	*10*	
	D	*10.00*	
	E	*11.51*	10 nmi = 11.51 mi
Unit to convert	**10**	*10*	
	E	*10.00*	
	D	*8.69*	10 mi = 8.69 nmi
Unit to convert	**10**	*10*	
	C	*10.00*	
	B	*32808.40*	10 km = 32808.4 ft

AV-22: RNAV FLIGHT PLANNING

This program is designed to be used by aircraft that are equipped with RNAV systems. It establishes the latitude and longitude parameters for waypoints which can then be used in rhumb-line navigation. Table 9-17 explains.

TABLE 9-17   PROGRAM AV-22: RNAV FLIGHT PLANNING

This program can be repeated between any pair of waypoints. A total of 43 waypoints may be established.

Data to be entered	Press key(s)	Display	Description of the display meaning
Identify program	2nd	0	
	Pgm	0	
	22	0	
Initialize program	SBR	0	
	CLR	0.0000	
Departure point	0	0	Waypoint 0 (Sante Fe, New Mexico)
Move to key position	A	0.0000	
Latitude	35.324	35.324	Latitude at Sante Fe in degrees, minutes, seconds
Move to key position	B	35.5444	Latitude in decimals
Longitude	106.039	106.039	Longitude at Sante Fe in degrees, minutes, seconds
Move to key position	C	106.0750	Longitude in decimals
Destination point	1	1	Waypoint 1 (Denver, Colorado)
Move to key position	A	1.0000	
Latitude	39.516	39.516	Latitude in Denver
Move to key position	B	39.8667	Latitutde in decimals
Longitude	104.451	104.451	Longitude in Denver
Move to key position	C	104.7528	Longitude in decimals
Halfway point	2	2	Waypoint 2 (Alamosa, Colorado)
Move to key position	A	2.0000	
Latitude	37.209	37.209	
Move to key position	B	37.3583	
Longitude	105.489	105.489	
Move to key position	C	105.8250	
	0	0	Waypoint 0
	D	0.0000	
	2	2	Waypoint 2

(Continued)

TABLE 9-17 (*Continued*)

Data to be entered	Press key(s)	Display	Description of the display meaning
	E	2.0000	
	2nd	2.0000	
	A'	6.3255	Rhumb-line course 006.3255° from Sante Fe to Alamosa
	2nd	6.3255	
	B'	109.5000	Distance from Sante Fe to Alamosa

AV-23: **TIME ZONE CONVERSIONS**

This program figures out the time anywhere on the globe, when given a certain time in a certain time zone. It does require that we know in which time zone a departure or destination point is located. (See Table 9-18.)

TABLE 9-18   PROGRAM AV-23: TIME ZONE CONVERSIONS

Data to be entered	Press key(s)	Display	Description of the display meaning
Identify program	**2nd**	*0*	
	Pgm	*0*	
	23	*0*	
Initialize program	**SBR**	*0*	
	CLR	*0.*	
Local time zone	**7**	*7*	Mountain standard time
Move to key position	**A**	*0.*	
Date	**11.29**	*11.29*	November 29
Move to key position	**B**	*0.*	
Local time	**15.0436**	*15.0436*	3:04:36 p.m.
Move to key position	**C**	*0.*	
Time difference (need not be entered. go to next step)	**0**	*0.*	+ if the sought after location is later, − if it is earlier than local time
Move to key position	**D**	*0.*	
Other time zone	**10**	*10*	
Move to key position	**A**	*0.*	
	E	*12.0436*	12:04:36 p.m. in Hawaii
	2nd	*12.0436*	
	E'	*11.29*	November 29

Since nearly all of the programs make use of the capability of the computer to place given values into the **A** through **E** and **A'** through **E'** key positions, a series of small plastic cards is provided, one with each program, which show what type of parameters are stored in each key position with each program. These cards are inserted into the upper slot of the TI59 (the only slot on the TI58C) whenever a certain program is used, and a handy carrying valet is provided to keep them always available. Figure 9-3 shows all of the cards, as well as some blank ones which can be filled in by the operator to be used with specially designed personal programs.

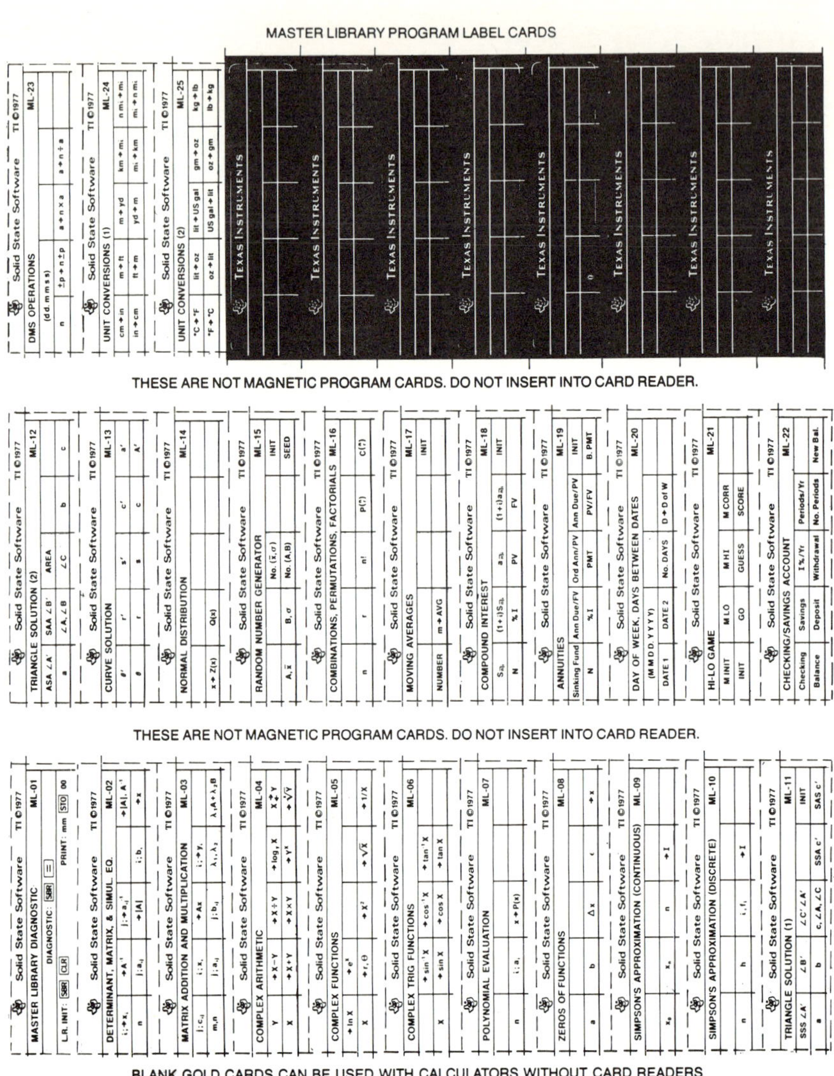

FIGURE 9-3 Parameter identification cards for the TI59 and TI58C. (Copyright 1977. Texas Instruments.)

10
HEWLETT-PACKARD'S PROGRAMMABLE HP-41C

The HP-41C is probably the ultimate in terms of small portable computers with applicability to aviation-related problems (see Figure 10-1). It's capabilities are so vast, that an entire book could be written about this computer alone and, in fact, the owner's handbook is 275 pages long. For our purposes here, we'll have to restrict our discussion of the HP-41C to a general description of its capabilities and of the ways in which it is operated for the purpose of solving problems related to the operation of airplanes.

The HP-41C is what is referred to as a programmable computer, meaning that the individual user can design and write his or her own programs whenever a given program is expected to be used with any degree of frequency. It is also a general-purpose computer which can be used to solve scientific, financial, and all manner of other specialized problems. The entire system consists of the computer itself, a card reader which will record special programs and then accept those programs in the form of magnetic strips for repeated use, and a printer which produces printed copies of all input and output (see Figure 10-2). The computer alone sells for around $375, the card reader costs $195, and the printer adds another $350, making the total for the system somewhere close to $1000.

The computer display provides a readout in alphanumeric form (letters and/or digits) in a liquid crystal presentation. The keyboard consists of 39 keys, 31 of which can be used to perform three different functions each; 5 keys are limited to one function, and the remaining 3 keys perform two functions. On the top of each key the basic function is printed in white. Above each key the second function is printed in yellow, and in order to cause any key to perform that second (yellow) function, the **TAN** function key must be pressed first each time.

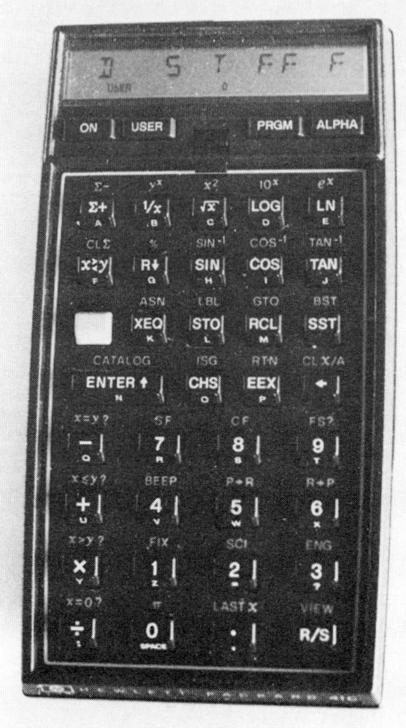

FIGURE 10-1 The Hewlett-Packard HP-41C programmable computer with aviation module.

On the slanted front edge of the keys, printed in blue, we find the letter (alpha) functions, representing the twenty-six letters of the alphabet plus =, ?, :, **SPACE**, and a comma. In order to have the keys operate in this mode, the **ALPHA** key must be pressed first, and when the keys are to be used in one of their other functions **ALPHA** must again be pressed to indicate that letters are no longer desired. Whenever a digit is needed as part of an alpha sequence, depressing the **TAN** key first will produce a digit rather than a letter.

In addition, any one of the keys, with the exception of the five function keys, can be programmed by the user to represent something other than what is printed on it if the operator desires to be able to key in certain values repeatedly by using only one key.

All of this, at first, appears incredibly complicated, and it takes a bit of time, practice, and study to use the various functions efficiently. As

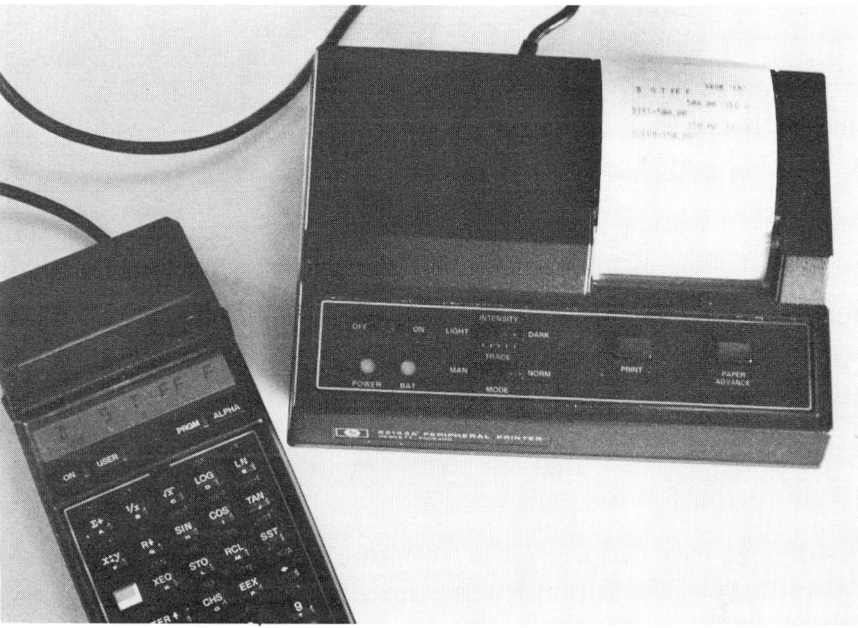

FIGURE 10-2   The HP-41C equipped with the optional card reader and the optional thermal printer.

appears to be true of all the computers, the instructions are written in a manner which the nonexpert will likely find difficult to comprehend.

The HP-41C "talks" back to the user, either asking questions, instructing the user what data must be entered next, or it will come back with such exasperating exclamations as *DATA ERROR, TRY AGAIN,* or *NONEXISTENT.*

In order to use the HP-41 in aviation, an aviation module must be plugged into it. (Hewlett-Packard offers some twenty such special-purpose modules.) This module contains the various aviation programs, and it comes along with a separate "Aviation Pac" instruction booklet which describes the aviation programs. These programs fall into six basic categories:

Flight Management   provides interchangeable solutions involving distance, speed, time, fuel flow, and fuel quantities. It can be used to determine ETE, time, distance and fuel to climb, endurance, range, and point of descent.

Weight and Balance   organizes the weight-and-balance calculations.

Flight Plan   produces a complete flight profile for up to six legs, including time and fuel during climb and descent, altitude changes en route, heading and ground speed based on known winds and, when used in conjunction with the weight-and-balance program, shows landing weight and center of gravity (CG).

Winds   consists of two programs which determine in-flight winds and determine head wind and crosswind components.

Position by One or Two VORs   determines the present position of the aircraft based on information obtained from the navigation receivers.

Mach Number and True Airspeed   converts calibrated airspeed to Mach and TAS.

Not included in the "Aviation Pac" are density altitude, area navigation programs, rhumb-line and great-circle navigation, and conversions. It would seem logical that these might be added at some future date in the form of a second aviation module.

One of the unique features of the HP-41C is that it has a nonvolatile memory, meaning that whatever was put into the computer last remains active in its memory even though it has been turned off. There are considerable advantages to this feature, though, at first at least, it tends to be annoying at times. It can be a bit disconcerting that whenever we turn the computer on it is likely to come up with some incomprehensible data which is left over from the last time we used it. The thing to do is to simply ignore it and start entering the new data which then automatically replaces the previous value or message. The good feature is that it eliminates the need to constantly reenter fixed values, such as moment arms (in weight-and-balance computations), TAS, fuel-flow figures, rates of climb, and other data which tends to remain the same as long as we are performing calculations for a given aircraft. Assuming that we have once established that our TAS is 150 knots on the average, at any time TAS comes up in any calculation, the computer will display *TAS=150?*, meaning that that is the figure it will use unless we decide to enter a different speed. The same is true of all the other values.

The keystroke sequences used in solving basic aviation problems (ETE, time to climb, fuel to climb, distance to climb, point of descent, endurance, range, and TAS) are described and illustrated in the individual chapters on each of these subjects, simplifying comparing the HP-41C with the other computers described in this book. The other programs are unique to the HP-41C and we'll now look at them in detail.

WEIGHT AND BALANCE AND FLIGHT PLAN

For some reason the WEIGHT AND BALANCE and FLIGHT PLAN programs always have to be used in conjunction with one another because there appears to be no way to tell the computer the amount of fuel on board before takeoff, except through the weight-and-balance sequence. Personally I think that that is kind of dumb, but that appears to be the way the program was designed. Instead of using a keystroke by keystroke illustration, I have here included the printout which the printer produced when I ran the program (see Figure 10-3). I did, in fact, run the program several times, changing a few of the values here and there until I came up with what I wanted.

The flight involves a pilot and one person in the right seat, but no other passengers. There are 100 pounds of baggage in the rear baggage compartment. The aircraft is fueled with 80 gallons of fuel. The flight is from Albuquerque (ABQ) via Santa Fe (SAF), Las Vegas, New Mexico (LVS), and Pueblo, Colorado (PUB), to Denver (DEN), consisting of four legs (see Figure 7-2). By giving the computer all of the known variables, it then goes ahead and computes all data applicable to every phase of the flight.

Here is how that works: We start by telling the computer the program we want to work, namely weight and balance. The computer already knows all the moment arms for each item, because they had been entered during previous weight-and-balance calculations. These, then, are the keystrokes: **ON XEQ** (execute) **ALPHA W B ALPHA** and now the display asks: *FWD BAG=0?* (is it correct that there is no baggage in the forward baggage compartment?). We agree that that is so and we press **R/S** (run/stop). The display now asks: *PILOTS=340?*

```
            XROM "WB"
FWD BAG=0.?
                    RUN
PILOTS=340.?
                    RUN
ROW1=0.?
                    RUN
ROW2=0.?
                    RUN
REAR BAG=100.?
                    RUN
FUEL-38.GAL?
            80.    RUN
GROSSWT=2,920.
                    RUN
NET WT=920.
                    RUN
MOM=259,195.
                    RUN
CG=89.
            XROM "PLAN"
VAR=-12.?
                    RUN
LEGS=4?
                    RUN
L1 CRS=043?
                    RUN
L1 DIST=47.?
                    RUN
FL1=7,500.?
                    RUN
L2 CRS=068?
                    RUN
L2 DIST=46.?
                    RUN
FL2=7,500.?
        9,500.    RUN
L3 CRS=358?
                    RUN
L3 DIST=162.?
                    RUN
FL3=9,500.?
        14,500.    RUN
L4 CRS=338?
                    RUN
L4 DIST=95.?
                    RUN
FL4=14,500.?
                    RUN
START ALT=5,352.?
                    RUN
DEST ALT=5,331.?
                    RUN


WA3=290.025?
                    RUN
WA6=300.040?
                    RUN
WA9=310.045?
                    RUN
WA12=315.055?
                    RUN
WA18=000.000?
                    RUN
CLIMB FUEL FLOW=28.0?
                    RUN
CLIMB TAS=120.0?
                    RUN
ROC=600.0?
                    RUN
CRUISE FUEL FLOW=22.0?
                    RUN
CRUISE TAS = 150.0?
                    RUN
DESCENT FUEL FLOW=12.0?
                    RUN
DESCENT TAS=150.0?
                    RUN
ROD=400.0?
                    RUN
T. O. TIME=?
            12.0000   RUN
LEG1
MC=043 D=47


CLIMB 5,352.-7,500.
HDG=025
GS=130.7
FUEL=1.7
REM FUEL=78.2GAL
LEG TIME=0:03:35
T/TIME=12:03:35

LEVEL AT 7,500.
HDG=028
GS=159.4
FUEL=5.4
REM  FUEL = 72.8 GAL.
LEG TIME= 0:14:46
T/TIME=12:18:20


LEG2
MC=068 D=46

CLIMB 7,500.-9.500
HDG=052
GS=145.0
FUEL=1.6
REM FUEL=71.2GAL
LEG TIME=0:03:20
T/TIME=12:21:40
LEVEL AT 9,500.
HDG=054
GS=175.1
FUEL=4.8
REM FUEL=66.4GAL
LEG TIME=0:13:00
T/TIME=12:34:40

LEG3
MC=358 D=162

CLIMB 9.500.-14,500.
HDG=336
GS=79.7
FUEL=3.9
REM FUEL=62.5GAL
LEG TIME=0:08:20
T/TIME=12:43:00

LEVEL   AT   14,500.
HDG=351
GS=122.9
FUEL=27.0
REM FUEL=35.5GAL
LEG TIME=1:13:40
T/TIME=13:56:40


LEG4
MC=338 D=95

LEVEL   AT   14,500.
HDG=335
GS=118.9
FUEL=9.9
REM FUEL=25.7GAL
LEG TIME=0:26:52
T/TIME=4:23:32

DESCENT 14,500.-5,331.
HDG=327
GS=109.3
FUEL=4.6
REM FUEL=21.1GAL
LEG TIME=0:22:55
T/TIME=14:46:28


LANDING WT, CG
GROSSWT=2,566.
CG=88.
```

FIGURE 10-3 The printout produced by the thermal printer when running a four-leg flight plan with the HP-41C.

(left and right seat occupants weigh 340 pounds?). Again we agree with that and press **R/S**. The display now wants to know: *ROW1=0?* (meaning, are there no passengers in the row 1 seats?). There are none so we again press **R/S**. The question now is: *ROW2=0?* and the answer is the same. **R/S** now produces: *REAR BAG 100?* (which the computer knew from the previous run of the program). We again agree and press **R/S**. Now the display asks: *FUEL=38.GAL?* (Here is where that memory comes in. That amount, 38 gallons, is what we had left after last running the program, and that's what the computer has remembered.) That is obviously the wrong figure so now we press **8 0** and the display changes to *80*, and then shows *GROSSWT=2,920*, the gross weight at takeoff. **R/S** then produces the net weight: *NET WT=920* (the computer already knew that the empty weight of the airplane was 2000 pounds), and it follows this by displaying the moment: *MOM=259,195* and the CG: *CG=89*.

So much for the weight-and-balance phase. Now, to work the flight-plan program we must again inform the computer of the desired program, by using these keystrokes: **XEQ ALPHA P L A N ALPHA** which results in the question: *VAR=−12.?* It wants to know if a −12° variation is correct for the area. This is needed because wind directions are given in true rather than magnetic degrees, and it will compute magnetic headings based on true wind directions. Since that figure is correct we press **R/S** and the computer asks: *LEGS=4?* Again that is correct (if we had previously run a program with a different number of legs, the computer would have used that figure instead) and **R/S** now produces: *L1 CRS=043?* (leg 1, magnetic course 043°?). This and most of the other figures were left in the computer memory from previous runs of this same program, so each time a correct value appears, we simply press **R/S** to look at the next: **R/S** : *L1 DIST=47.?* (leg 1 distance 47 nautical miles?); **R/S** : *FL1=7,500?* (first flight level 7500 feet msl?); **R/S** : *L2 CRS=068*; *L2 DIST=46.?*; *FL2=7500?* This is not acceptable because the terrain between Santa Fe and Las Vegas is quite high. So we now punch in: **9 5 0 0** and the display changes to *9,500*. **R/S** : *L3 CRS=358?*; **R/S** : *L3 DIST=162.?*; **R/S** : *FL3=9,500.?* Again we don't like that altitude, because with the winds coming over the mountains, it'll be a lot smoother at a higher altitude. So we now punch in: **1 4 5 0 0** and the display changes to *14,500*. **R/S** : *L4 CRS=338?*; **R/S** : *L4 DIST=95.?*; **R/S** : *FL4=14,500?*; **R/S** : *START ALT=5,352?* Now, having received all the navigation information the computer wants to know the elevation of the departure airport which, in this case, is 5352 feet. **R/S** : *DEST ALT=5,331?* It now wants to know the elevation at the destination airport, Denver, which is 5331 feet. **R/S**:*WA3=290.025?* It is now asking for the winds at 3000 feet. Since we won't be flying at 3000 feet, the figure is meaningless and the computer will automatically ignore it. **R/S** : *WA6=300.040?* It now wants to verify that the winds at 6000 feet are blowing from 300° at 40 knots. This will affect our flight only during brief periods after takeoff and before landing, but the computer will take that into account. **R/S** : *WA9=310.045*; **R/S** : *WA12=315.055*; **R/S**:*WA18=000.000*. As they are not reported, the computer does not give the winds at 15,000 feet, and since we are not planning to go to 18,000 feet, that figure does not matter. **R/S**: *CLIMB FUEL FLOW=28.0?* Whenever there are more than ten digits or letters in a display, they move across from right to left, leaving the last ten alphanumerics in the display. **R/S** : *CLIMB TAS=120.0?*; **R/S** : *ROC=600.0?* Are we anticipating a rate of climb of 600 feet per minute (ft/min)? Yes. **R/S** : *CRUISE FUEL FLOW=22.0?*; **R/S** : *CRUISE TAS=150.0?*; **R/S** : *DESCENT FUEL FLOW=12.0?*; **R/S** : *DESCENT TAS=150.0?*; **R/S** : *ROD=400.0?* (rate of descent, 400 ft/min?); **R/S** : *T.O. TIME=?* The computer now wants to know when we plan to take off. Since we want to get going at noon, we punch in **1 2 . 0 0 0 0** and the display shows *12.0000*. We now have entered everything that the computer needs to know. We now press **R/S** and the display shows a little birdlike figure, vaguely resembling a constipated seagull, which moves haltingly across the display to show that the computer is busy thinking. It takes a few seconds but when it comes back to life it starts to present the entire flight in orderly succession without any need for additional keystrokes. The information is itemized by legs, showing first the magnetic course (*MC*) for each leg and the distance (*D*) for each. It then shows the altitude change to be accomplished as part of that leg (*CLIMB* or *DESCENT*), followed by the magnetic heading during climb (*HDG*), the ground speed during climb (*GS*), the fuel used in climb (*FUEL*), the fuel remaining upon reaching the new altitude (*REM FUEL*), the time in climb (*LEG TIME*), and the actual time at which the new altitude will be reached (*T/TIME*). It then tells us that we're level at the new altitude and follows that with all the data applicable to the rest of the first leg which is flown at the given cruising altitude. By the time all the data for each leg has been presented, it will tell us when we arrive at our destination, and it will give us the landing weight and landing CG (*LANDING WT*; *CG*; *GROSSWT*; *CG*).

The whole process takes a lot less time than it does to describe it, and whenever we plan a long flight and we are uncertain about the best altitude in terms of either time en route or fuel used, we could easily run this same program for several different altitudes and then compare the results.

POSITION BY ONE OR TWO VORs

In order to determine the position of the aircraft relative to either one or two VORs, the computer contains two programs, labeled 1VOR and 2VOR. I'm not certain that either one of these is of much use, but just to show how they work, I've included the printout from one run of the 1VOR program (see Figure 10-4).

To activate the program we first have to tell the computer what we're after by using these keystrokes: **ON XEQ ALPHA TAN** function key **1 V O R ALPHA** and the computer now wants to know the wind direction and velocity: *WIND=334.032?* (direction 334°, velocity 32 knots).

```
            XROM "1VOR"
WIND=334.032?
                     RUN
VAR=12.?
                     RUN
MH=327.
            023.     RUN
TAS=150.?
                     RUN
T AT R1=?
        12.0000      RUN
R2<DEG>=?
            336.     RUN
R2<DEG>=?
            068.     RUN
T AT R2=?
        13.1624      RUN
DIST-150.NM
```

FIGURE 10-4 The thermal printer hard copy resulting from running the POSI-TION WITH ONE VOR program.

If that is correct we can go on. If it is not, we enter the applicable data. The next question refers to magnetic variation: *VAR=12?* We accept that and go on by pressing **R/S** which results in the display of the last magnetic heading which we used in a previous operation: *MH=327*. Since we're currently holding a heading of 023°, we enter that value and then press **R/S**. The resulting question is: *TAS=150.?* We accept that and go on to the next question: *T AT R1=?* (time at the first radial from the VOR). We now enter the time in hours, minutes, and seconds (12:00:00 in our example), and the computer now wants to know the radial 1 direction in degrees: *R1⟨DEG⟩=?* The direction being 336° we enter that value, and are now asked for the direction in degrees of the second radial: *R2⟨DEG⟩=?* For it we enter 068. Now we are asked for the time when we find ourselves on top of that second radial: *T AT R2=?* We look at our clock and enter 13:16:24. Now the computer has all the data it needs and after a moment it comes up with the result: *DIST=150NM*, meaning that we are currently 150 nautical miles from the VOR on the 068° radial.

I frankly can't imagine anyone actually using this program in flight, because the only possible use it would have would be in the event that the pilot is not sure of his position, in which case he or she is likely to be too nervous and upset to provide the computer with all that data. Similarly the 2VOR program seems useless, because anyone who can't locate him- or herself when two VORs are being received shouldn't be up there in the first place.

Beech Aircraft Corporation has recently introduced an especially programmed aviation module for the HP-41C which contains in its memory all the performance parameters and limitations of the Super King Air 200. It is identified as the Beechcraft SKA 200 FPC Custom Application Module (SKA stands for *Super King Air* and FPC stands for *Flight Planning Computer*). While it can be used to perform all preflight and in-flight calculations for pilots flying Super King Airs, its most interesting feature is the so-called SAVE program, which is an effective fuel-conservation feature. It takes into account forecast winds aloft, fuel required to climb to, cruise at, and descend from each legal altitude for any given flight, and provides the pilot with a readout of the total fuel and total time required for the flight at both the most economical and the shortest en-route-time altitudes, and it also displays the savings of fuel and the increases in time which can be expected if the most economical altitude is selected.

The module also includes the necessary programs to compute rhumb-line and great-circle navigation, which are missing in the standard aviation package.

Since it is not especially complicated for someone familiar with the techniques involved in computer programming and with the performance parameters and limitations of specific aircraft to design such a special program, it would appear logical that similar special-purpose modules will eventually be available for any number of high-performance aircraft.

I might add here that the printer is relatively compact and lightweight, and it can function off its batteries without the need for any external power source (though continued use in this fashion will reduce the life of those batteries). This means that the printer along with the computer can be taken along in the aircraft and can be used in flight.

11
INSTRUCTION BOOKLETS AND OPERATING MANUALS

Manufacturers of highly technical products seem to be under the impression that the best person to write the operating manual is the one who was primarily responsible for the design of the product. This is a fallacy, as can be seen by virtually all operating manuals provided in conjunction with portable aviation computers as well as those which are supposed to explain the workings of avionics systems.

Because the writers of these manuals are thoroughly familiar with the workings and peculiarities, they always take a great deal of technical knowledge on the part of the operator for granted. They insist on using phrases and abbreviations which may be part and parcel of the language of computer experts, but which cause uninitiated readers to shake their heads in dismay.

I am not a computer expert. I am simply the average pilot with a lot of hours (5000) in the air, whose aviation activity is such that I would be a logical customer for many of the products which are discussed in this book. I hate to admit it, but in some instances I have had to read those operating manuals from cover to cover three or four times in order to figure out what the author had in mind, and even then it often took many hours of working with the actual hardware to get the hang of it.

What I am attempting to do in the following chapters is to look at various portable computers from the point of view of the user, the average pilot or aircraft owner who pays good money for these products but who, quite often, may be unable to make full use of their capabilities, because he or she may have neither the time nor the patience to gain full understanding of the product from the manual.

In the back of the book I have included a glossary of terms and abbreviations which includes not only those which are actually used in the book, but also those which are found with any degree of frequency in the various manuals.

12
EVALUATION AND ANALYSIS

How good are all these little electronic marvels in terms of usefulness in our everyday flying activity? There can be no simple answer to this question, just as there can be no easy determination that one model is superior to all others. Too much depends on the type of flying we do and the type of airplane we do it in.

Table 12-1 provides a quick comparison of the types of functions which are available with each of the systems discussed in the foregoing and following chapters. For the purposes of this evaluation let us eliminate the E6b and the Mathematician (which is used here as representative of all low-cost electronic calculators, no matter the manufacturer) and just look at the six models which are designed specifically for use in aviation.

First there is the cost. Prices range from just under $70 to $375, not including, in the cases of the two most expensive units, the optional printers. Thus, if spending as little money as possible is an important consideration, the Avstar obviously wins hands down.

The second row shows the number of keys comprising the different keyboards. At first glance it may not make a great deal of difference how many keys there are, as long as they are capable of activating the functions needed to solve aviation problems. But that is misleading. Too many keys tend to become confusing, especially if we anticipate using the computer during actual flight when turbulence and other cockpit distractions may make it difficult to always hit the right key at the right moment. Also, in each case some or many of these keys are designed to perform several different functions, and since few of us can ever expect to be able to actually remember all these functions, and the means by which they are called up, it is important that the identification of these functions is clearly shown.

TABLE 12-1 CHARACTERISTIC FEATURES OF VARIOUS AVIATION COMPUTER MODELS

Feature	E6b Flight Computer	Mathematician	Avstar	N60 Navigator	Heathkit OC-1401	Navtronic 1701tr	TI58C/TI59	HP-41C
Price in 1981 $'s	5–25	15+	70	100	150	240	300	375
Number of keys		32	40	60	36	30	45	35
Display limit (digits)		8	8	10	9	8	10	10
Display type*		L	C	L	L	L	L	C
Addition		●	●	●	●	●	●	●
Subtraction		●	●	●	●	●	●	●
Multiplication		●	●	●	●	●	●	●
Division		●	●	●	●	●	●	●
Change sign		●	●	●	●	●	●	●
Floating decimal		●	●	●	●	●	●	●
Higher math functions		●		●	●		●	●
Memory (number)		1	1	2	9	0	60	62
Logic†		P	A	A	P	A	A	P
Talk back							●	●
Error indication			●	●	●	●	●	●

* L = LEDs; C = liquid crystals.

† A = arithmetic logic; P = Polish logic.

Here, again, the Avstar appears rather well designed. The basic (primary) functions of each key are printed in black on a silver background and can be easily seen in any halfways decent light. The secondary functions, primarily dealing with conversions, are printed in smaller type and in blue and tend to be hard to see. But then they are rarely used in flight. Punching the keys requires a degree of precision because, in view of the small size of the computer, they are fairly close together and one might easily hit the wrong key by accident.

TABLE 12-1 (*Continued*)

	E6b Flight Computer	Mathematician	Avstar	N60 Navigator	Heathkit OC-1401	Navtronic 1701tr	TI58C/TI59	HP-41C
Timer function; counting up					●	●		
Timer function; counting down						●		
Fuel remaining; constant display					●			
Convert time to decimals			●	●	●	●	●	●
Convert decimals to time			●	●	●	●	●	●
Nautical to statute miles and vice versa	●		●	●	●	●	●	
°F to °C and vice versa	●		●	●	●	●	●	
Kilometers to nautical miles and vice versa	‡		●	●	●		●	
Pounds (U.S.) to kilograms and vice versa	‡		●				●	
Meters to feet and vice versa			●	●			●	
Gallons to pounds and vice versa							●	
Gallons to liters and vice versa			●		●		●	
Write-in programs							●	●
Nonaviation programs available							●	●
Can be used to program nav. avionics							§	●

‡ Available on some models.

§ TI59 only.

While the Avstar does not actually tell the operator what information is needed in order to perform a given computation, the keys are outlined in such a fashion that it serves as a reminder to the pilot to enter all the necessary parameters. Next to the more expensive Navtronic 1701, it might be said that the Avstar is the easiest to use for the average general-aviation pilot, flying singles or light twins. It is also the only one which fits easily into the average shirt pocket.

The Commodore N60 Navigator with its 60 keys capable of performing a total of 99 functions, is not only physically the largest but also (to me, at least) the most confusing of the group. One of its confusing features is the fact that many keys have inscriptions which are so similar, that one always has to think twice to figure out which key is the one that needs to be pressed at a given moment. For example, there are keys marked **DIST**, **DIST1**, **DIST2**, **DIST FLN**, **DIST OFF**, each of which is used to accomplish a different purpose. And, even though the different groups of keys are in varying colors, it takes a lot of practice to remember what to do in which order.

The Heathkit OC-1401 suffers from a similar problem. Of its 36 keys, 34 perform dual functions and in many instances triple functions. While the descriptions are color coded to simplify using the right function key, it does take a bit of getting used to. Also, this computer, too, has instances in which 2 keys are using the same identification, and the only difference between them is the color of the background on which the description is printed. The keys themselves are spaced sufficiently far apart to make hitting the right key easy, and keystrokes produce a definite click which I find reassuring. As is true of the other two models discussed so far, the pilot must remember the information needed by the computer to perform a given function and, in some instances, the consecutive order in which this information must be entered.

The Navtronic 1701 uses the least number of keys and is also the least confusing of all of the computers discussed here. Its most unique feature, the ability to tell the pilot what information it wants next in order to do its job of computing, makes it extremely easy to use. Here, too, the keys are placed at a reasonable distance from one another, so that hitting the right key is no great problem.

The TI58C and TI59 can be examined along with the HP-41C, both being highly sophisticated computer systems designed to accept customized programs in addition to the (optional) built-in aviation programs. The Texas Instruments models use a 45-key keyboard with all but 3 keys performing multiple functions, while the Hewlett-Packard model uses 35 keys, most capable of performing at least three functions, to accomplish more or less the same thing. Despite their relatively small size, the keys are well spaced and operation is easy.

The third row identifies the number of digits which can be accommodated by the display. They vary from 8 to 10, the number being of no particular importance to the user as all are ample to provide a display of the kind of information we are usually after. One primary difference in the type of display is the way in which time is shown when expressed in hours:minutes:seconds. The Avstar and the N60 use dashes (*12-30-30*), the Navtronic uses spaces (*12 30 30*), the OC-1401 uses periods (*12.30.30*), but the TI58C, TI59, and the HP-41C use just one period (*12.3030*) which tends to get confusing at times, because one is never certain whether time value is expressed in terms of hours:minutes:seconds or decimals.

The fourth row talks about the type of display. An LED (light-emitting diodes) display refers to internally lit digits (all in red except for the N60, in which they are green) which can be read in the dark but tend to be hard to see in bright sunlight. The liquid crystal displays used in the Avstar and the HP-41C are black numbers on a grey background which can be seen in any adequate light but not in the dark. LED displays require a considerable amount of electrical current, and all units equipped with them use rechargeable nicad batteries, requiring connection with an external power source (house current or aircraft or automobile current) using a converter, in order to be recharged or used for a long period of time. The current drain associated with liquid crystal displays is so small that rechargeable batteries are not needed. (The manufacturers of the Avstar claim that it will operate on one set of its tiny batteries for up to 1000 hours.)

The fifth through eleventh rows deal with the calculator portion of the computers. Here all have more or less the same capabilities, though the Avstar is not designed to perform higher-mathematics functions.

The twelfth row presents the number of memories in which data can be stored. This is a great convenience feature and the Navtronic is the only one in the group with no memory availability. The Avstar has 1, the N60 has 2, the Heathkit has 9, and the other two have 60 and 62, respectively. In the Texas Instruments and HP-41C computers large numbers of memories are needed to permit performance of some of the more complicated and lengthy computations.

The thirteenth row identifies the type of logic used. The OC-1401 and the HP-41C use Polish logic while all others use arithmetic logic. This is strictly a matter of personal preference. Both perform perfectly well, as soon as one gets used to using either one or the other.

Talk back, the subject of the fourteenth row, refers to the ability of the computer to tell the operator what data it needs in which order in order to perform the desired computation. The Navtronic 1701 performs this function by causing little red lights to light up next to a label which describes the needed data. The HP-41C goes a giant step further. It provides a question in the form of actual written words in the display, showing the previously used value for a given parameter, followed by a question mark. If that value is acceptable to the operator, he or she can go on and use it. If not, the operator can change it to a new value.

All models use some type of error indication when unacceptable data are accidentally entered (shown in the fifteenth row). In some the word *ERROR* appears in the display. In others the display simply starts to flash, indicating the error. The HP-41C has a variety of such indications. It will display either the word *ERROR* or *TRY AGAIN* or *NONEXISTENT*, depending on the type of mistake involved.

The sixteenth through eighteenth rows deal with timer functions. Only two of the models include a timer, the Navtronic and the Heathkit. I, personally, find this function very useful and if I opted for the purchase of a Navtronic, I would never consider one without that capability. The Navtronic permits counting time either up or down, and in the latter case, when *00 00 00* is reached, an audible alarm begins to beep, making it most useful for timed instrument approaches. The timer function cannot be used to keep track of fuel used or fuel remaining, except that time indications can, at intervals, be multiplied by fuel flow to display the amount of fuel used so far. The timer in the Heathkit will not count down in the time format, and it has no alarm feature. On the other hand, it can be used to display fuel remaining on a continuing basis, assuming that fuel on board and fuel flow were entered before the start of the flight.

The nineteenth through twenty-seventh rows deal with the different types of conversions which these computers will perform.

The twenty-eighth row, labeled "Write-in programs," indicates whether the computer is designed to accept customized programs, and the twenty-ninth row, labeled "Non-aviation programs available," indicates whether these same computers can accept modules for use in financial, scientific, technical, or other specialized fields. The last row shows which computers can be programmed to execute complex navigation avionics programs.

So there we have it. If the airplane you're flying is a cabin twin, turboprop, or jet, then my suggestion would be that either of the Texas Instruments TI58C or TI59 or the Hewlett-Packard HP-41C are the right computers for you. With fuel costs being what they are today, these computers will soon pay for themselves in increased economy of operation. If the airplane is of a lesser kind, then one of the other models will probably do the trick just as well.

The question, of course, remains: Do we need any of these gadgets in the first place? After all, people have been flying by the seat of the pants, by guesswork, or any other nonscientific method for years. So what's all the fuss about?

Quite honestly, these gadgets are nice to have, but they are not a necessity. But, using them and doing so regularly will tend to help us understand the capabilities of our airplanes to a greater degree. They will cause us to become interested in flying with greater precision and, in view of the increasing complexity of today's airspace, that can only help to increase the utility and economy of the airplane.

PART TWO

BASIC AVIATION CALCULATIONS

AVIATION CALCULATIONS PERFORMED BY VARIOUS COMPUTER MODELS

Problem to solve	Data needed	E6b	Mathematician	Avstar	N60 Navigator	Heathkit OC-1401	Navtronic 1701tr	TI58C/TI59	HP-41C
Hours to decimals	Hours:minutes:seconds			X	X	X	X	X	X
Time in decimals to h:min:s	Hours and decimals of hours			X	X	X	X	X	X
Estimated time en route	Distance; ground speed	X	X	X	X	X	X	X	X
Time to climb	Altitude change; rate of climb	X	X	X	X	X	X	X	X
Fuel to climb	Time to climb; fuel flow	X	X	X	X	X	X	X	X
Distance to climb	Time to climb; ground speed	X	X	X	X	X	X	X	X
Point of descent	Altitude change; rate of descent; ground speed	X	X	X	X	X	X	X	X
Fuel used in descent	Time of descent; fuel flow	X	X	X	X	X	X	X	X
Time and fuel from takeoff to landing	Climb time, distance, fuel; descent time, distance, fuel; total distance, cruise fuel flow		X	X	X	X	X	X	X
Endurance	Fuel on board; fuel flow	X	X	X	X	X	X	X	X
Range	Endurance; ground speed	X	X	X	X	X	X	X	X
Density altitude	Pressure altitude; temperature	X		X	X		X	X	
True airspeed	Pressure altitude; temperature; IAS or CAS	X		X	X	X	X	X	X
Ground speed and heading with wind	TAS; course; wind components			X	X		X		
Wind direction and velocity in flight	Course; heading; ground speed; TAS			X	X		X	X	X
Rate of climb (gradient, distance)	Altitude change; distance; ground speed		X	X	X	X	X	X	
Area Navigation	Two points relative to one VOR				X		X		
	Two points relative to two VORs						X		
Rhumb-line navigation	Latitude, longitude (two points)						X	X	X
Great-circle navigation	Latitude, longitude (two points)							X	
Standard temperature	Altitude			X					
Timer counting up	Takeoff time					X	X		
Elapsed time, counting up	0:00:00					X	X		
Timer counting down	Time to critical point						X		
Fuel remaining	Fuel on board; fuel flow					X			
Timed approaches	Distance; ground speed						X		
DME slant-range correction	Altitude; station elevation; radial			X				X	
Weight and balance	Moment arms; weights		X	X	X	X	X	X	X
Line of sight distance	Altitude; terrain and station elevation							X	
Present position (dead reckoning)	Point of departure; heading; ground speed; time							X	
Lowest usable flight level	Barometric pressure							X	
TAS versus Mach	TAS or Mach; density altitude							X	
Turn performance (stall, G force, diameter)	TAS, angle of bank							X	
Freezing level	Altitude; temperature					X		X	
Distance to station	Ground speed; time	X	X	X	X	X	X	X	X
Time to station	Ground speed; distance	X	X	X	X	X	X	X	X

13

CONVERTING HOURS : MINUTES : SECONDS TO DECIMALS AND VICE VERSA

In aviation we are constantly faced with time values expressed in decimals which, for better understanding, we would like to see expressed in hours:minutes:seconds. And then there are times when the opposite is called for. Using a standard calculator this can be a fairly complicated problem, involving a lot of steps if we are dealing with minutes as well as seconds. On the other hand, many of the portable aviation computers can perform this task in the flick of an eye.

As is obvious from Figure 13-1, there are a lot of steps involved when we're using a standard calculator. For this example I decided to start with 3:47:34 which I then transformed into decimal values. The first thing to do is to temporarily ignore the hours (3) as that figure is not going to change. So we start by dividing the number of minutes (47) by 60 and the display shows that they represent *.78333333* hours. But now, since the next calculation is designed to turn the seconds into decimals of a minute, we'll multiply the last result by 100 which produces *78.333333*. This we stash temporarily in the memory while we undertake the next portion of the problem. We enter the seconds (34) and divide that figure by 60 and the display now shows *.56666666*. We now retrieve the stored value from the memory and add it to the above, producing a display showing *78.899999*. Now we have to reverse the process where we multiplied by 100. We therefore now divide that figure by 100 which turns the display into *.78899999*. Now we add the 3 (hours) which we had ignored earlier, and the result appears as *3.7889999*, which represents the decimal equivalent of 3:47:34.

Now, in order to accomplish the reverse, we start with 3.7889999 and again get rid of the hours by deducting the 3, resulting in a display

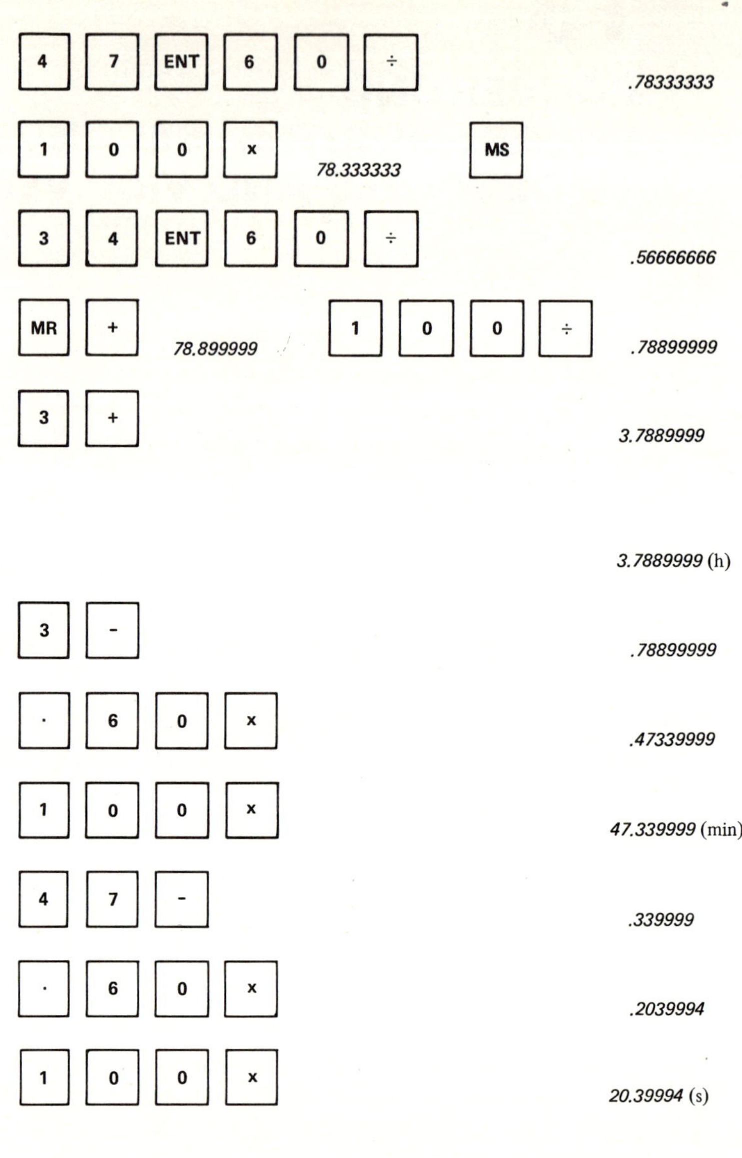

FIGURE 13-1   Using an ordinary calculator to convert time values is a tedious process.

of *.7889999*. We multiply that by 0.60 which produces *.4733999* which we then multiply by 100 resulting in a display of *47.339999* which represents the minutes. We now get rid of the minutes by deducting 47 which leaves *.339999* and we multiply that by 0.60 and get *.2039994*. That is now multiplied by 100, and the display changes to *20.39994*, representing the seconds. Thus the total result shows *3:47:20*. It appears that somewhere along the line, probably due to the rounding off of fractions in the calculator, we have lost 14 seconds. Oh, well.

Now, using the Avstar, the entire procedure is quite simple (see Figure 13-2). Here I have taken 2:37:54 as the time value to be changed into decimals. We simply enter the time in hours:minutes: seconds which produces a display showing *2-37-54* and then press the plus key (**+**), then the zero key (**0**), and then the equal key (**=**) and the display shows *2.6316667* as the decimal value which equals 2:37:54.

To accomplish the opposite, we enter the decimal value in rounded-off form (2.63) and then press plus (**+**), then **T:** and then equal (**=**) and the display shows *2-37-48*. Again we have lost a few seconds due to the rounding off of fractions.

The N60 uses just about the same format as the other computers (see Figure 13-3). To change hours:minutes:seconds to decimals, the keystrokes are **(inv) F 0**, while to accomplish the opposite, they simply are **F 0**.

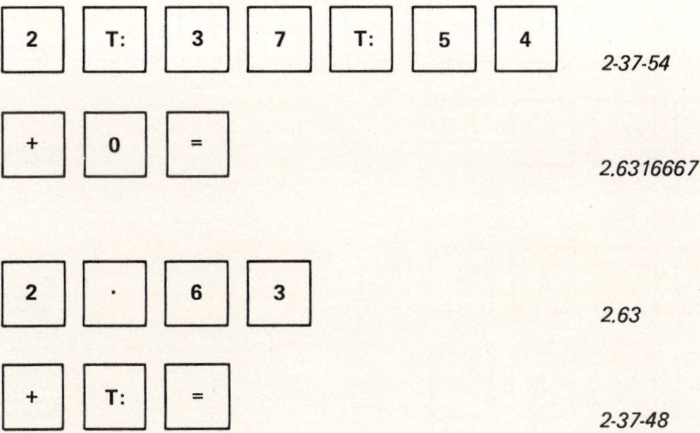

FIGURE 13-2   With the Avstar time conversions require three keystrokes.

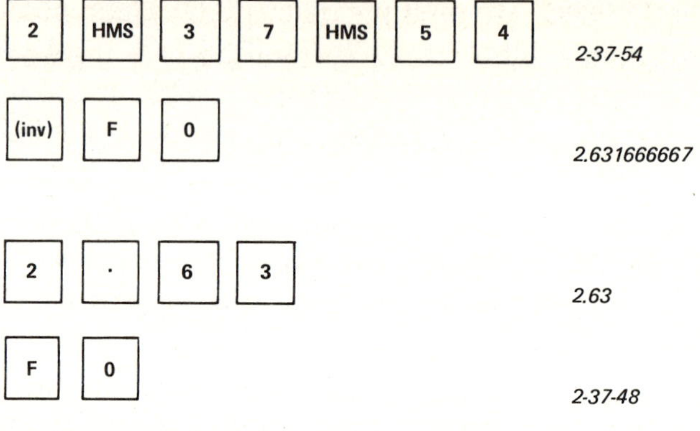

FIGURE 13-3 The N60 uses either two or three key-strokes to convert time.

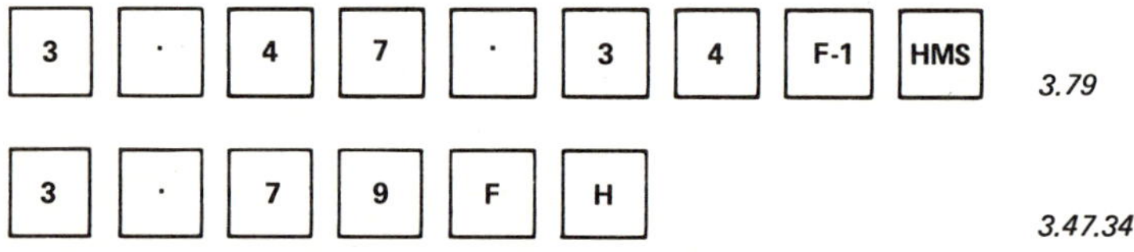

FIGURE 13-4 Time conversions take two keystrokes with the OC-1401.

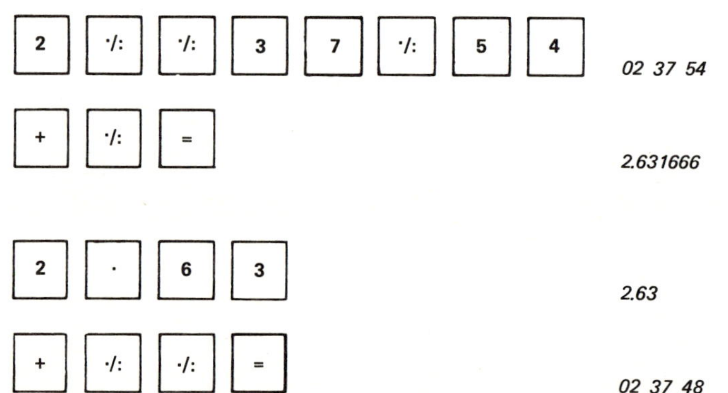

FIGURE 13-5 The Navtronic 1701 requires either three or four keystrokes for time conversions.

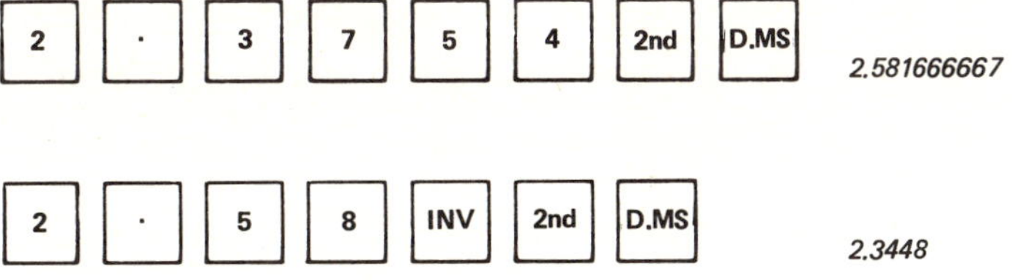

FIGURE 13-6   The TI59 uses two or three keystrokes to perform time conversions.

Converting time values from decimals to hours : minutes : seconds and vice versa is incredibly simple with the OC-1401 (see Figure 13-4). To accomplish the former, we simply press **F H**. And to accomplish the latter we press **F-1 HMS** instead. That's all.

Using the Navtronic, the routine is quite similar, though the key strokes are a bit different (see Figure 13-5). Again we start by entering the time in hours, minutes, and seconds and follow that by **+**, then **./:**, and then **=** and the display will show *2.631666*, the decimal value for 2:37:54.

To do the reverse we again enter 2.63 followed by **+**, then **./:** twice, and then **=** and the display shows *02 37 48*. Apparently the two computers are in cahoots in that both lost 6 seconds somewhere along the way.

To accomplish these conversions with the TI59 (58C) involves either two or three keystrokes (see Figure 13-6). Pressing **2nd D.MS** will convert hours : minutes : seconds, entered as 2.3754 (meaning 2:37:54) to hours and decimals of hours: *2.581666667*. To accomplish the opposite, we press **INV 2nd D.MS** and 2.58 will show up as *2.3448* or 2 hours, 34 minutes, and 48 seconds. This computer, too, apparently loses 6 seconds in the process. Maybe that's some kind of a congenital defect.

14
ESTIMATED TIME EN ROUTE (ETE)

Of the various basic aviation problems, this is probably the one most often encountered. Before virtually any flight we will want to know how long it is going to take us to get to our destination, and we'll usually start out by taking a rough guess such as, if the distance to be covered is 750 nautical miles, and our airplane usually cruises at a TAS of 150 knots, then it doesn't take a great deal of brain power to figure out that it will take 5 hours to get to where we want to go. But when the figures involved aren't quite that obvious, and if we'd like to have a somewhat more precise estimate than would be possible by figuring things in our head, then any one of the units which we have discussed will give us the desired answer.

In the example shown in the following illustrations, I have arbitrarily chosen a distance of 245 nautical miles, and a ground speed of 150 knots. In actual fact we won't know what our ground speed is going to be until we get into the air and find out what the wind components are which we have to deal with, so in the initial stage of planning we are likely to use the TAS figure as the expected ground speed.

Figure 14-1 shows the position of the E6b which provides the required answer. The indicator has been placed opposite **15** which, in this instance stands for the ground speed or 150. Then, by reading the time value on the inner circle which appears opposite the distance which is shown as **24**, meaning 240 nautical miles and **25**, meaning 250 nautical miles, we find that the time en route will be a bit less than 1 hour and 40 minutes (**1:40**). This may not be terribly precise, but it is certainly good enough at this stage in the game.

In Figure 14-2 we have the steps involved in solving the problem by using an ordinary calculator. We know that the mathematical formula

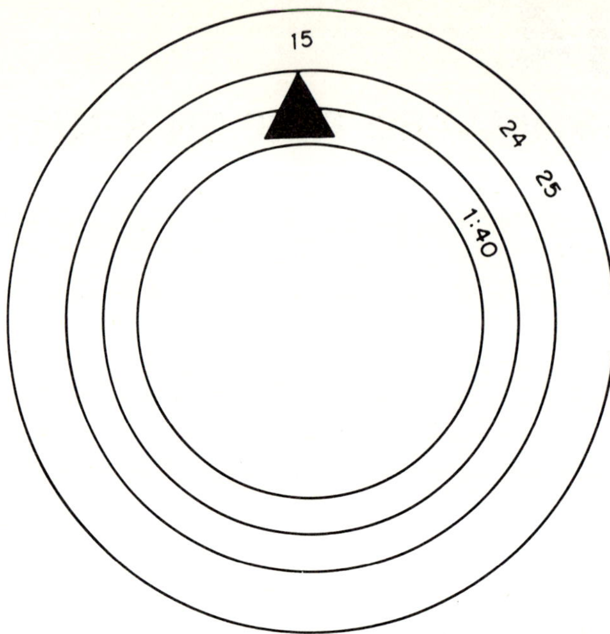

FIGURE 14-1 Figuring ETE using the E6b.

calls for dividing the distance by the speed. We therefore enter the distance (245) and then enter the speed (150) and by pressing the ÷ key, we command the calculator to divide the first value by the second. The display now shows *1.6333333*. In other words, we're looking at 1 hour and a little more than 0.6 of an hour. We now want to convert that fraction into minutes. To accomplish this we deduct 1 in order to get rid of the 1 hour, and the display now shows *.6333333*. We now punch in 60 and command the calculator to multiply, and the

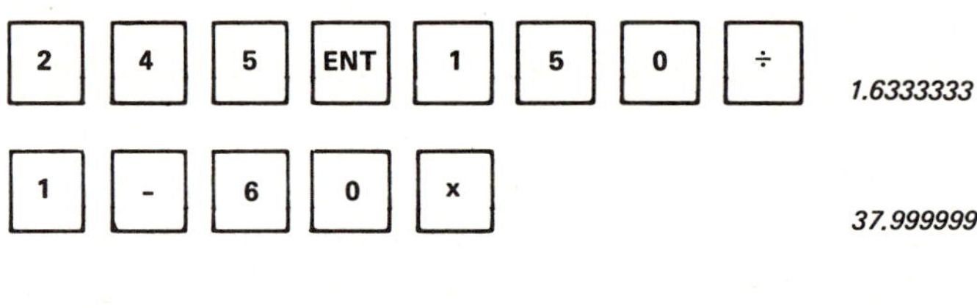

FIGURE 14-2 Figuring ETE using an ordinary calculator, this one using Polish logic.

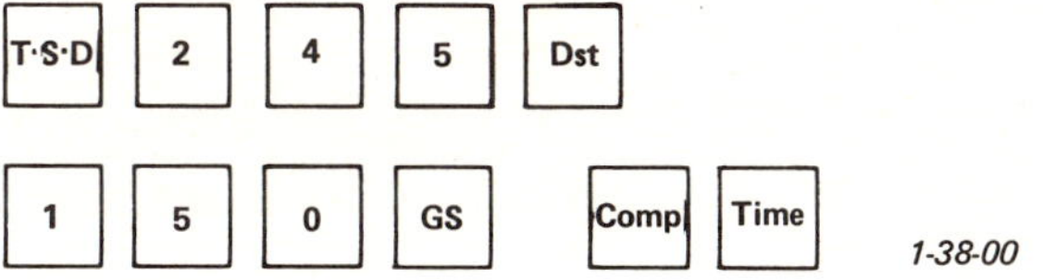

FIGURE 14-3   The Avstar uses its T-S-D program to calculate ETE.

display will show *37.999999* which, for all practical purposes, represents 38 minutes. So, by adding the 1 hour back again, we now know that the time en route figures out to 1 hour and 38 minutes (*1:38*).

In Figure 14-3 we show the steps involved in using the Avstar. Here we first have to tell the computer that the problem to be solved falls into the time-speed-distance category. We therefore start by pressing the **T-S-D** key. We now enter the distance (245) and, by pressing the **Dst** key we let the computer know that this value is the distance involved. We then enter the speed figure (150) and identify its meaning by pressing the **GS** key which stands for *ground speed*. We then command the computer to start computing by pressing the **Comp** key followed by the **Time** key which tells the computer that the answer we want is a time value. The display now reads *1-38-00* or 1 hour and 38 minutes.

The Commodore N60 Navigator provides the answer by the simple process of division, plus the conversion of the result from decimals into hours, minutes, and seconds (see Figure 14-4). We start by entering the distance (245) then use the ÷ key and enter the speed (150) followed by = and the result is displayed as *1.633333333*. We now press **F** (the function key) followed by the **0** key (actually the **(d)dms** key, but zero is easier to remember) and the display changes to *1-38-00*.

The Heathkit OC-1401 contains no special program for this pur-

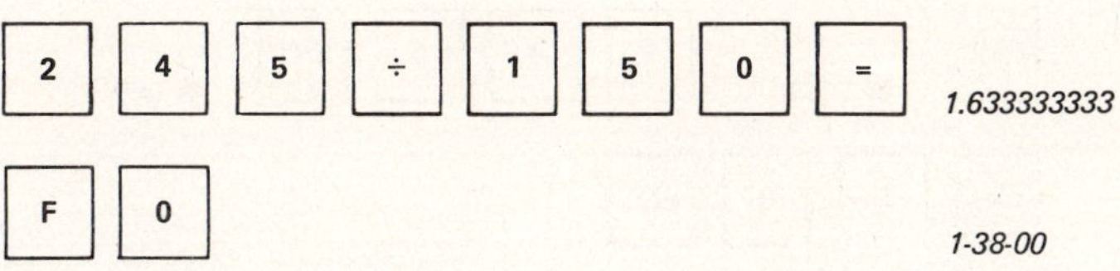

FIGURE 14-4   The N60 has no special program. It uses calculator functions to figure ETE.

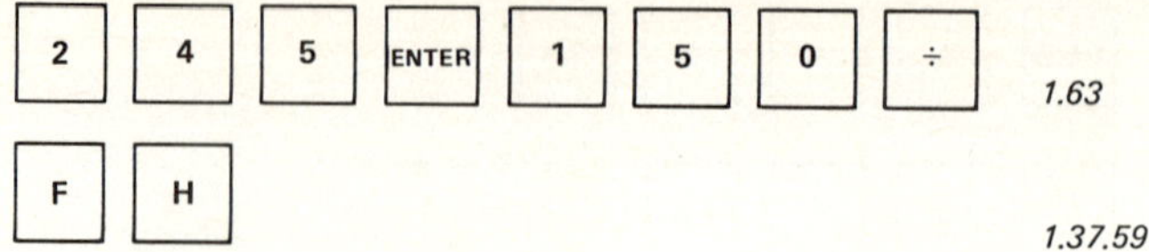

FIGURE 14-5   The Heathkit OC-1401, like the N60, uses calculator operation to figure ETE.

pose. We simply use the calculator portion of the computer to divide the distance (245) by the expected speed (150) and the display will show *1.63* (see Figure 14-5). We now press **F H** and the display changes to *1.37.59* or 1 hour, 37 minutes, and 59 seconds. One of the nice features of this unit is that it doesn't bother us with a half dozen or more decimal figures which only tend to be confusing. The computer display is designed to remain constantly in the two-decimal position, though in its innards it actually takes account of all the remaining fractional values which are not displayed.

In Figure 14-6 we are using the Navtronic 1701tr aviation computer. In this case we start out by pressing the **TIME** key in order to tell the computer the type of value we're after. The computer instantly responds by activating the little red light next to the word **DIST**, which tells us that the computer now wants to know the distance involved. We use the numerical keys to enter the distance (245) and press **ENT** to place that value into the computer memory. By activating the red light next to **SPD** the computer now asks for the speed figure. We punch in 150 and press **ENT** and without delay the display will show *01 38 00* or, again, 1 hour, 38 minutes, and no seconds.

With the TI59 (58C), two of the programs stored in the aviation module can be used. One is program number 02 (FLIGHT PLAN WITH WIND). We start out by calling up that particular program

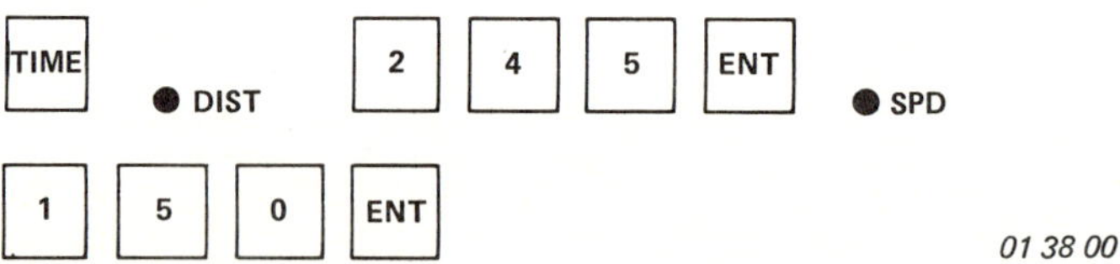

FIGURE 14-6   The Navtronic 1701 uses the built-in TIME program for ETE calculations.

with these keystrokes: **2nd Pgm 0 2**. Then, in order to accomplish what is referred to in the instructions as "initializing the program" we press **SBR CLR** which stands for *subroutine clear*, and eliminates any stray data that may be still kicking around in the computer as the result of previous use. We then enter the distance (245) and assign it to key **B** by pressing that key. We then enter the TAS (150) and assign it to **D'** by pressing **2nd D'**. We now press **D R/S** (run/stop) and the display will show *1.3760* which stands for 1 hour, 37 minutes, and 60 seconds or, in fact, 1 hour and 38 minutes (see Figure 14-7*a*). It is a peculiarity of this computer that it occasionally comes up with 60 seconds instead of raising the number of minutes by one. What we have done here is to use only a small portion of a much longer program which was discussed in Chapter 9.

Since calculating the ETE with no allowance for wind is one of the simplest of calculations, it would be easier and faster to simply divide the distance by the speed which would produce a readout of

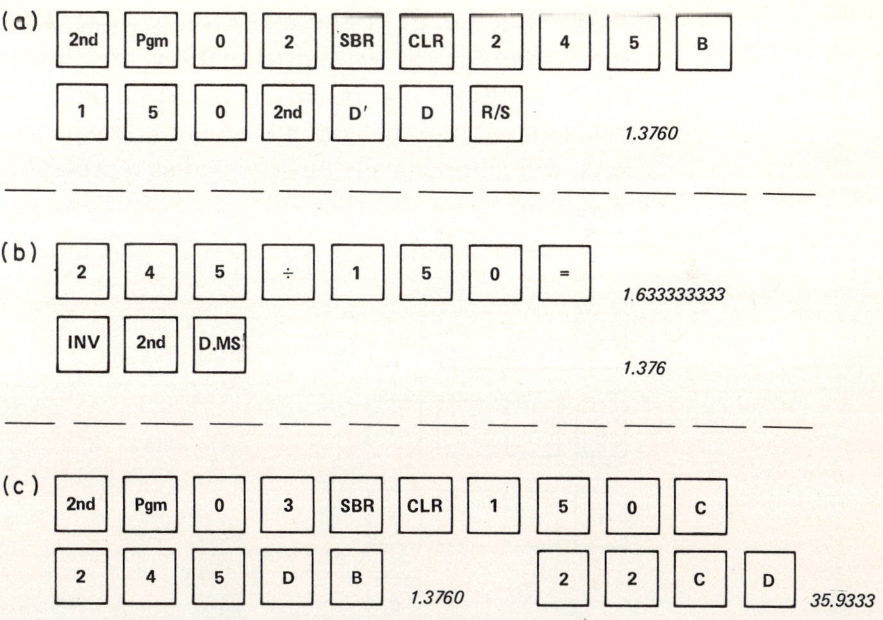

FIGURE 14-7   The TI59 has three processes available to figure ETE. The first one (a), using the AV-02 program, results in the time value in hours:minutes:seconds. The second one (b) in the illustration uses the calculator function. The third (c) uses the AV-03 program producing time values in hours:minutes:seconds, and, assuming fuel flow has been entered, it also shows fuel used en route.

1.633333333. Then, by pressing **INV 2nd D.MS** we arrive at *1.376* which is again 1 hour, 37 minutes, and 60 seconds (see Figure 14-7*b*). (**D.MS** stands for *degrees*, *minutes*, and *seconds* and is used for degrees as well as time.)

Another alternative is to use the AV-03 program which, in addition to providing the answer to the time question, will also tell us the amount of fuel needed, assuming the fuel-flow rate is entered (see Figure 14-7*c*). I have assumed a fuel flow of 22 gal/h.

When using the Hewlett-Packard HP-41C solving an ETE problem is part of the FLIGHT MANAGEMENT program, which is stored in the Aviation Module. In order to activate this program we must first press **XEQ** (execute) followed by **ALPHA** which changes the meanings of the keys from their mathematical to their alphabetic functions (see Figure 14-8). We then press **F** (for flight) and **M** (for management) followed by again pressing **ALPHA** to return the keys to their normal functions. The display now shows *D S T FF F* which stands for *distance*, *speed*, *time*, *fuel flow*, and *fuel amount*. Each letter is positioned above a key in the top row, the *D* (distance) letter above the **A** key, the *S* (speed) letter above the **B** key, the *T* (time) letter above the **C** key, the *FF* (fuel flow) letters above the **D** key, and the *F* (fuel amount) above the **E** key. From now on in, as long as we are using this program, any data entered into the computer followed by pressing one of the above keys, will automatically be accepted as representing whatever the key stands for.

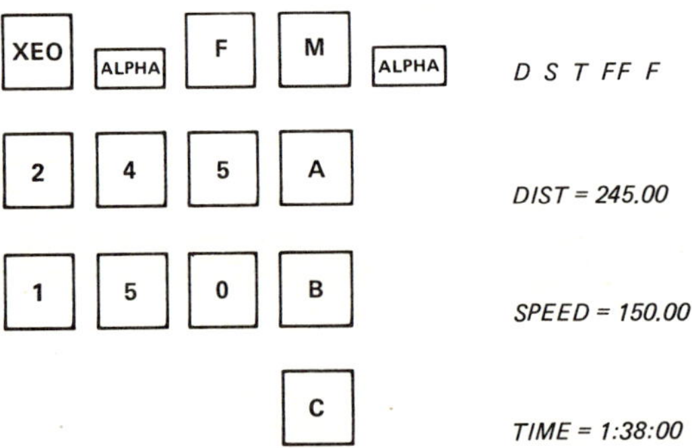

FIGURE 14-8 The HP-41C uses an all-purpose FLIGHT MANAGEMENT program to figure ETE and a whole lot of other aviation calculations.

So now we press 245, the distance, followed by using the **A** key to identify the meaning of that figure, and the display now appears as *DIST=245.00*. Next we enter the speed (150) followed by the **B** key to identify it to the computer and the display now reads *SPEED=150.00*. We now press **C** to tell the computer that we want to know the time and the display changes to *TIME=1:38:00*.

15
TIME TO CLIMB

Whenever there is reason for us to climb to a relatively high altitude, we're likely to want to know how long it will take to get there. For this purpose we must first deduct the airport elevation from the eventual cruising altitude in order to find the altitude change, the number of feet which we actually have to climb. Then we divide that figure by the number of feet per minute which we figure we'll be averaging on the way up. The resultant figure represents the number of minutes which it will take to get to our cruising altitude.

In the accompanying illustrations we have assumed an airport elevation of 2528 feet, a desired cruising altitude of 11,500 feet, and a rate of climb averaging 600 ft/min.

If we want to use the E6b for this purpose we must first deduct the airport elevation from the desired cruising altitude in our head (or on paper). A quick guess would place the amount of altitude change at about 9000 feet, which is close enough for our purposes. We now place the indicator which is also marked with the number **60** opposite **10**, the 60 in this case representing 600 ft/min, and the 10 representing 1 minute (see Figure 15-1). Then, by checking the relationship of the figures on the first and second ring, we find that **90**, representing 9000 feet, is close to opposite **15**, representing 15 minutes. In other words, it will take us just a bit less than 15 minutes to climb to 11,500 feet msl.

Using the calculator, everything is really quite simple: We enter the desired cruising altitude (11,500) and deduct the airport elevation (2528) and the display will show *8972* which represents the number of feet we will have to climb (see Figure 15-2). We now divide that by 600 and the display will show *14.95333*. If we now wanted to we could convert that decimal portion of the figure into seconds, but being as

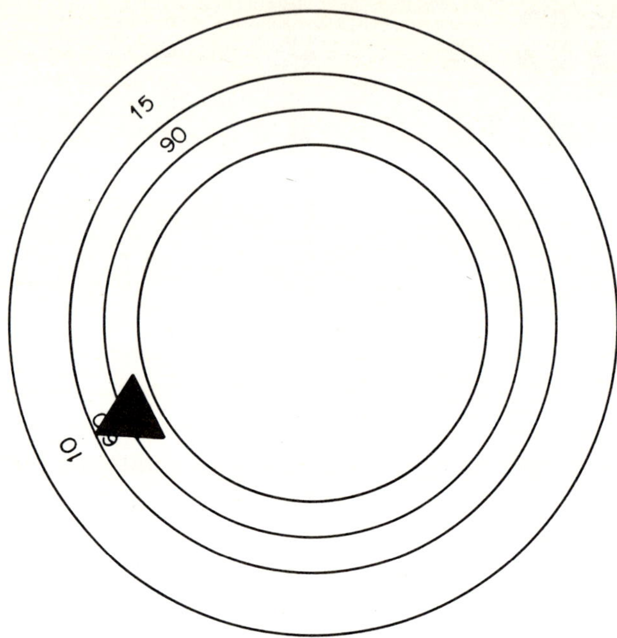

FIGURE 15-1   Figuring time to climb, using the E6b.

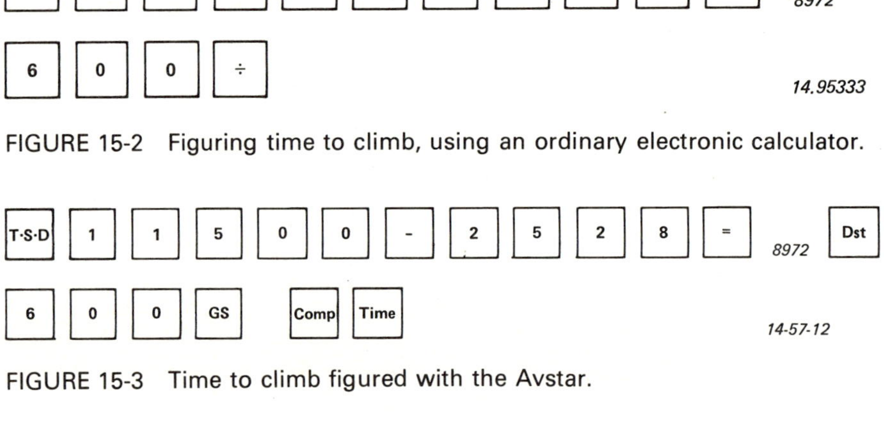

FIGURE 15-2   Figuring time to climb, using an ordinary electronic calculator.

FIGURE 15-3   Time to climb figured with the Avstar.

FIGURE 15-4   Time to climb figured with the N60.

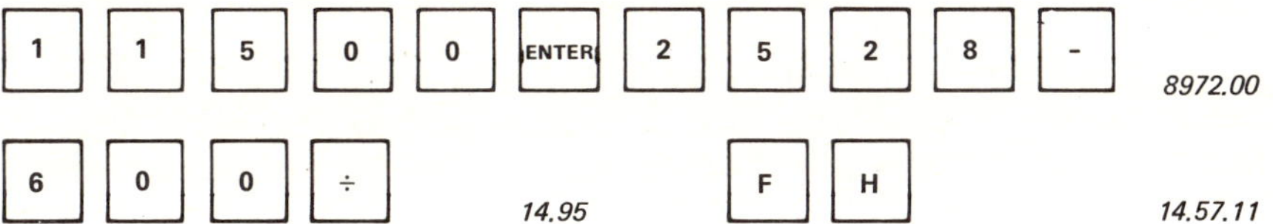

FIGURE 15-5   Here the OC-1401 is used to figure time to climb.

high a figure as it is, there seems to be little point in it. (It would figure out to 57.1998 seconds.) The answer, quite obviously, is extremely close to 15 minutes.

Using the Avstar, the routine is actually a bit more complicated (see Figure 15-3). First we call up the appropriate program by pressing **T-S-D** and then we punch in the cruise altitude (11,500) and deduct the airport elevation (2528) which produces a display of *8972*. We now press the **Dst** key to identify that value as vertical distance. We then punch in 600 and press **GS** which in this instance represents vertical rather than horizontal speed. Then, by pressing **Comp** and **Time**, the display window will respond with *14-57-12* or 14 minutes, 57 seconds, and $^{12}/_{60}$ of a second.

The Commodore N60 again employs simple arithmetic steps plus the conversion of the decimal result to hours, minutes, and seconds (see Figure 15-4). We enter the desired cruise altitude (11,500 feet) followed by − and followed by the departure-airport elevation (2528) followed by = and the displayed result is *8972*. We now press ÷ to divide the altitude difference by the rate of climb, 600, followed by = and the result reads *14.95333333*, the time in minutes and decimals of minutes. We now press **F** followed by **0** and the display changes to *14-57-12* which, in this case, again means 14 minutes, 57 seconds, and $^{12}/_{60}$ of a second.

The OC-1401 also uses simple calculator technique (see Figure 15-5). We deduct the airport elevation (2528) from the desired flight altitude (11,500) and divide the result (8972) by the rate of climb (600) to arrive at *14.95* minutes. Then, by pressing **F H** the display changes to *14.57.11* or 14 minutes, 57 seconds, and $^{11}/_{60}$ of a second.

With the Navtronic 1701tr the routine is quite similar, except that, again, the computer talks back to us by asking for the next information which it needs in order to solve the problem (see Figure 15-6). We first press **TIME** and the red light asks for **DIST**. We now enter the cruise altitude (11,500) and deduct the airport elevation (2528) and the

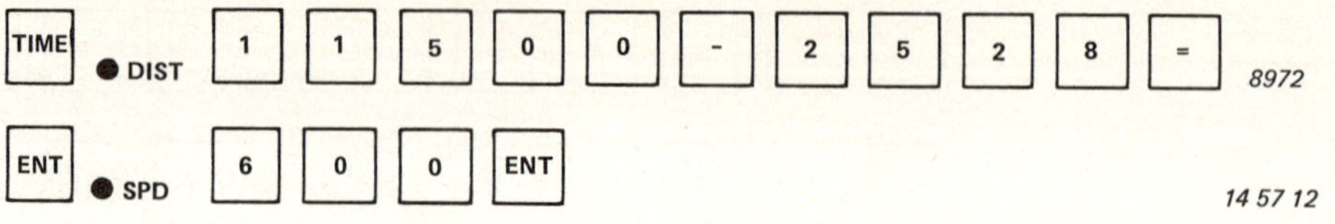

FIGURE 15-6   One of the Navtronic 1701 programs can be used to figure time to climb.

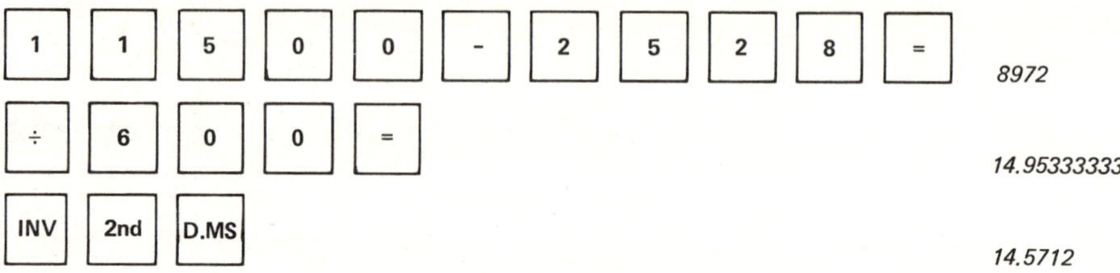

FIGURE 15-7   Time-to-climb calculations using the TI59.

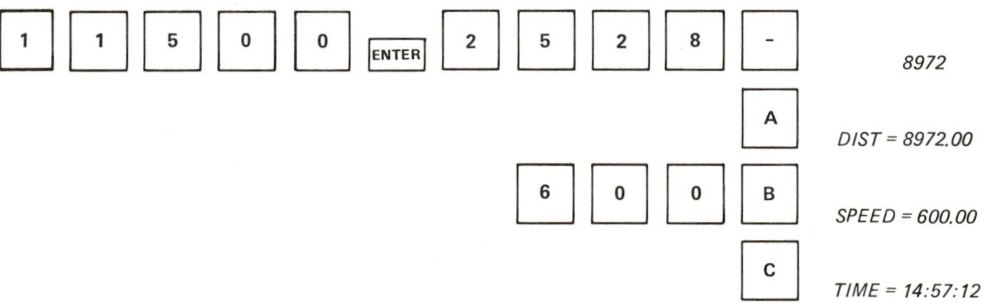

FIGURE 15-8   With the HP-41C it is again the FLIGHT MANAGEMENT program which is used to figure time to climb.

display will again show *8972*. We punch **ENT** and now the red light asks for **SPD**. We punch in 600 and press **ENT** and the display will respond with *14 57 12* (14 minutes, 57 seconds, and $^{12}/_{60}$ of a second).

This, too, is a simple calculation and there is no specific program for it in the aviation programs for the TI59 (58C). We deduct the airport elevation from the desired cruising altitude to arrive at the amount of altitude change and divide that by the rate of climb. In our example the display shows *14.95333333* (see Figure 15-7). That is then converted to a time value by pressing the three necessary keys, and the display now shows *14.5712* or 14 minutes and 57 seconds (and $^{12}/_{60}$ of a second) as the time to climb.

In order to provide the desired result using the Hewlett-Packard HP-41C, the computer is still in the FLIGHT MANAGEMENT program mode which was described in detail when we solved the ETE problem. All we do now is enter the degree of altitude change which, representing vertical distance, is given to the computer as distance (see Figure 15-8). We press 11,500 (the desired cruise altitude) followed by **ENTER** (because the computer uses Polish logic) and follow it by inserting the airport elevation (2528) followed by using the minus key (−) and the display shows *8972*. We now press **A** to identify that figure as distance and the display changes to *DIST=8972.00.* We now enter the rate of climb (600) and press **B** to identify that figure as the vertical speed. The display now shows *SPEED=600.00.* We now press **C** in order to get a result in terms of time, and the display now shows *TIME=14:57:12.* (If we had accidentally pressed the **D** or **E** key, the computer, since it insists on remembering what had been entered into it previously, would have displayed either *F FLOW=* whatever figure had been used previously for fuel flow, or *FUEL=* and the previous entry for the amount of fuel either on board or remaining.)

16
FUEL TO CLIMB

Now that we know how long it is going to take us to get to cruising altitude, the next thing we'd probably like to know is how much fuel are we going to burn on the way up. To arrive at that figure, what has to be done is to multiply the fuel flow per hour in terms of gallons per hour or pounds per hour by the amount of time it takes to get up there. But since we have been talking in terms of minutes when figuring the time to climb and since fuel flow is figured in terms of hours, we must either convert the fuel flow into gallons or pounds per minute and then multiply, or we must convert the minutes and seconds into a comparable fraction of an hour and then multiply the gallons per hour (or pounds per hour) by that figure.

When using the E6b there is no need to go through all those mathematical gyrations (see Figure 16-1). We simply place the indicator opposite the fuel flow (28) and then, opposite the figure on the second ring which represents the minutes in climb, we can then read the fuel burned (70, which in this instance stands for 7 gallons). It's really quite simple, assuming that we can remember which figure stands for what.

With the calculator it gets a bit involved (see Figure 16-2). Let's assume that we still have that last time figure (14.95333 minutes) in the display, we now place that in memory by pressing **MS**. We then enter 60, recall the figure from memory by pressing **MR**, and call for divide by pressing the ÷ key. The display will now show *4.0124841* which represents the fraction of an hour. We again place that in the memory by pressing **MS**. We then punch in the gallons per hour figure (28), recall the previous value from memory by pressing **MR**, and again ask the calculator to divide. The display will show *6.9782207*, meaning that we'll be using just under 7 gallons to get to cruising altitude.

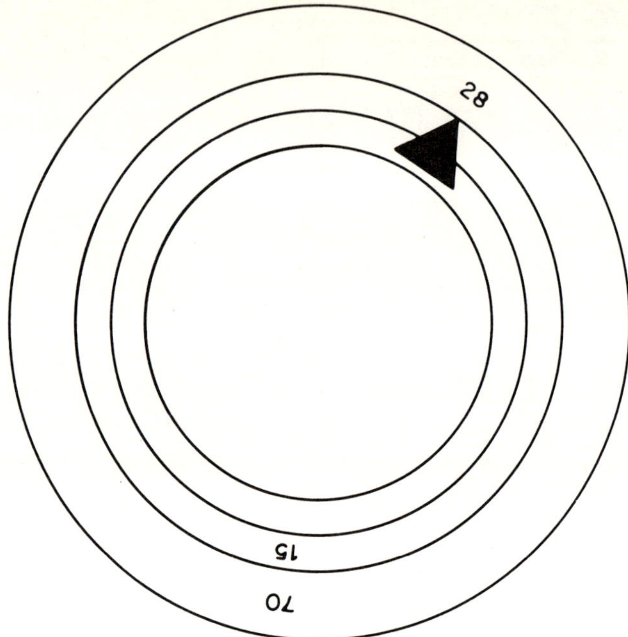

FIGURE 16-1   Fuel to climb, figured with the E6b.

Using the Avstar, things are a bit simpler, but it does require that we remember how it works. For some reason which is not necessarily immediately clear, the problem falls under the time-speed-distance category. So we press **T-S-D** and follow it by entering the minutes and seconds (**T:**, **14**, **T:**, **57**) followed by identifying that value to the computer by pressing **Time** (see Figure 16-3). The display now shows *0-14-57*. We now punch in the gallons per hour figure (28) and identify it as fuel flow by pressing the **FPH** key. That is followed by using the

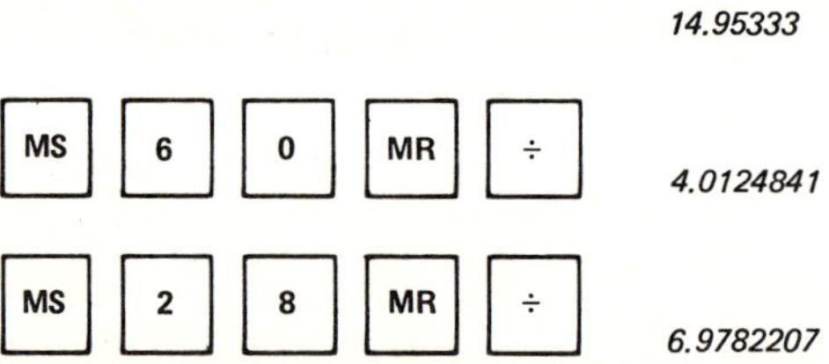

FIGURE 16-2   Using a standard calculator to figure fuel to climb.

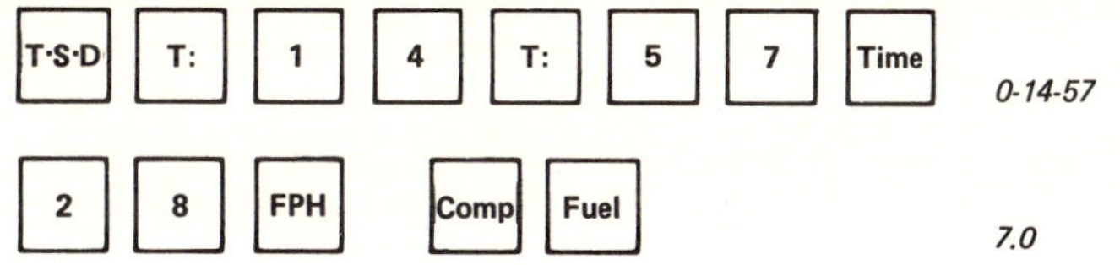

FIGURE 16-3   Fuel-to-climb calculations use the T-S-D program in the Avstar.

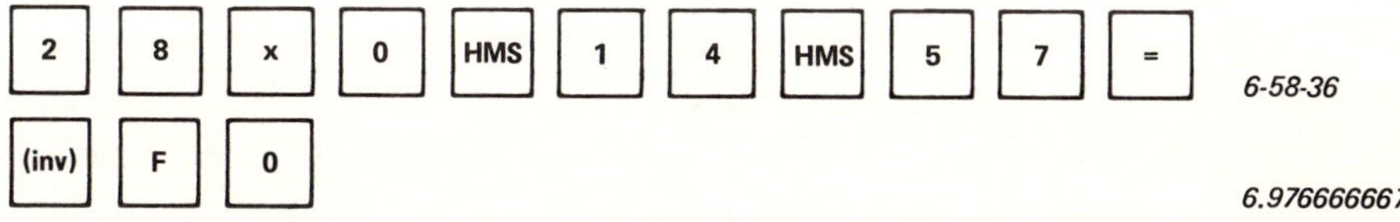

FIGURE 16-4   Fuel to climb, using the N60.

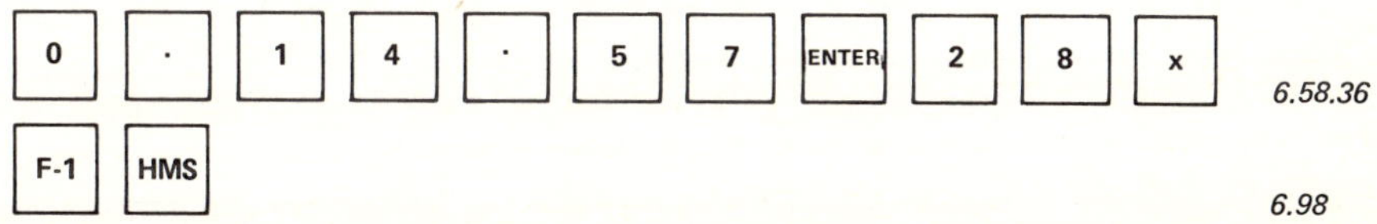

FIGURE 16-5   Fuel to climb, figured with the OC-1401.

Comp key and the **Fuel** key and the display will come up with *7.0*, the number of gallons burned during climb.

The Commodore N60 permits us to simply multiply the fuel flow by the time spent in climb and then to convert the result, which initially shows up as a meaningless time figure, into a decimal figure which then represents the amount of fuel used during climb (see Figure 16-4). We start by entering the fuel flow (28) followed by the multiply key (**×**) and then followed by the time in hours, minutes, and seconds: **0 HMS 14 HMS 57**, and follow that with **=**, resulting in a display showing *6-58-36*. We now press **(inv)** and **F** plus **0** and the display changes to *6.976666667*, the amount of fuel needed for the climb.

Remembering the time to climb we enter it into the OC-1401, this time in hours, minutes, and seconds (0.14.57) and multiply that figure by the fuel flow used in climb (28) and the resulting display will show up in the hours : minutes : seconds format (see Figure 16-5). To convert that to a figure which is representative of the number of gallons used during climb, we press **F-1 HMS** and the display now changes to *6.98*.

With the Navtronic 1701tr the problem requires even less thought and actions by the operator (see Figure 16-6). We start by pressing **FUEL** because that's what we are interested in. The red light now commands us to provide the value for **RATE**, meaning the fuel-flow figure. We punch in 28 followed by **ENT** and the red light asks for **TIME**. To enter 14 minutes and 57 seconds we use these keystrokes: **./: ./: 1 4 ./: 5 7** followed by **ENT** and the display will come up with **7.0**.

In Figure 16-7 we use the previously obtained time value, 0:14:57 by entering 0.1457 into the TI59 (58C) and convert it to decimals by pressing **2nd D.MS** and the display shows *.2491666667*. We multiply that by the fuel flow and the display now shows the fuel used during climb: *6.976666667* gallons.

The HP-41C is still in the FLIGHT MANAGEMENT program mode which has been used all along. We enter the time in hours,

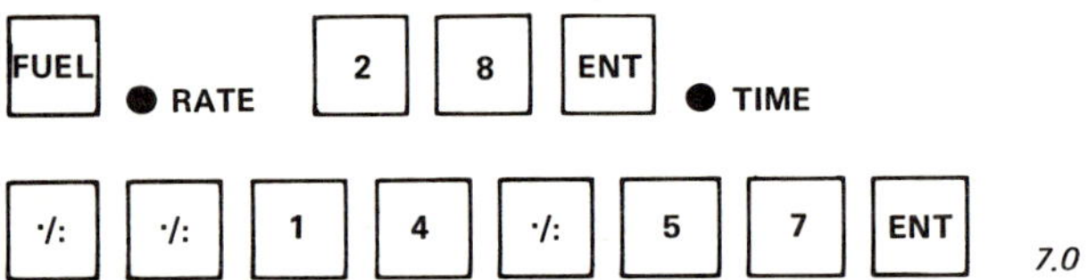

FIGURE 16-6   The Navtronic 1701 figures fuel to climb by using one of its programs.

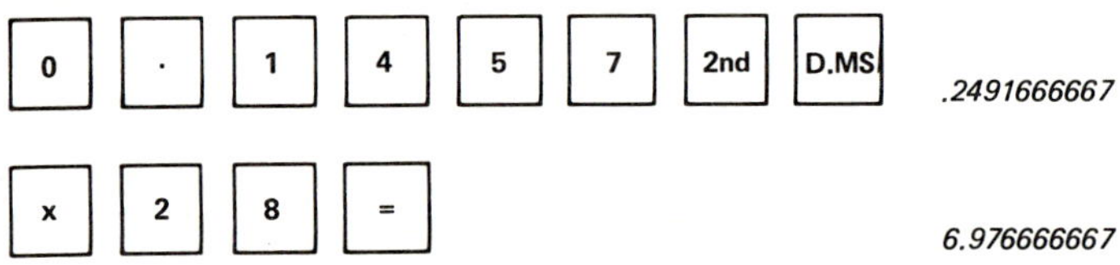

FIGURE 16-7   The TI59 multiplies time by fuel flow.

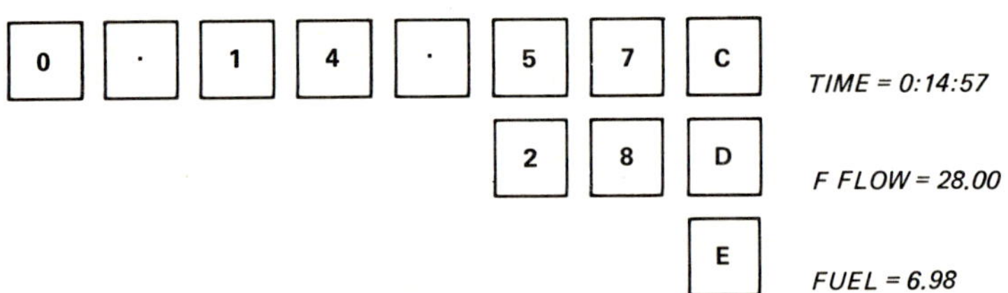

FIGURE 16-8   With the HP-41C the FLIGHT MANAGEMENT program is again employed.

minutes, and seconds by pressing **0** . **14** . **57** and the **C** key and the display shows *TIME=0 : 14 : 57* (see Figure 16-8). We now press 28 (the fuel flow) followed by the **D** key and the display changes to *F FLOW=28.00*. We now press **E** which stands for the amount of fuel and the display changes to *FUEL=6.98*. Once we get used to using the FLIGHT MANAGEMENT program in the HP-41C, all the basic time, distance, speed, and fuel problems can be solved with ease.

17
DISTANCE TO CLIMB

Now, as long as we're on the subject of climbing to altitude, we might just like to know how much horizontal distance we'll be covering during climb. What we need to know in order to determine that distance, is the average ground speed during climb and the time to climb. Well, we have previously established the time to climb as just under 15 minutes, and for the purpose of this exercise, we'll assume that the ground speed during climb will be 120 knots.

When using the E6b we place the indicator opposite the expected average ground speed (**12**, which this time stands for 120 knots), and then we find the time in minutes (**15**) on the second ring, and opposite it on the outer ring we'll find the number **30**, which means that we'll be covering 30 nautical miles horizontally during climb (see Figure 17-1).

When using the average calculator we start by entering our ground speed (120) and dividing it by 60 which produces a display showing *2* (see Figure 17-2). We then enter the time figure in decimals which was left over from the previous calculations (14.95333) and call for multiplication and the display will come up with *29.90666* or just under 30 nautical miles.

When using the Avstar, this is again a time-speed-distance problem, so we start by pressing **T-S-D** (see Figure 17-3). We then enter the ground speed (120) followed by **GS** to identify the meaning of that figure. Next we enter the time (14 minutes, 57 seconds), followed by pressing **Time** to identify that value, then followed by **Comp** and **Dst** as the command to the computer to compute the distance value. The display will now read *29.9*. I might point out that we could just as well have entered the time value with its identification first, and the

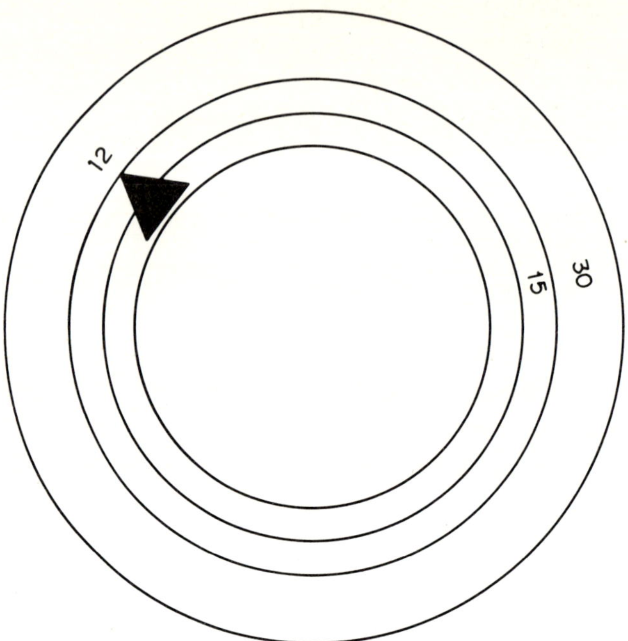

FIGURE 17-1   Distance to climb figured with the E6b.

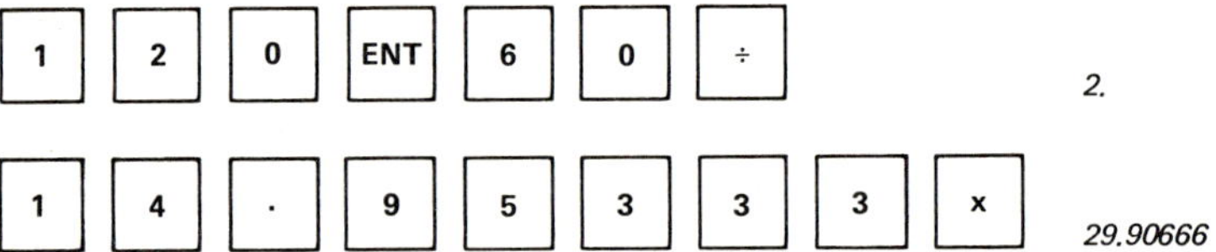

FIGURE 17-2   Distance to climb figured with an ordinary calculator.

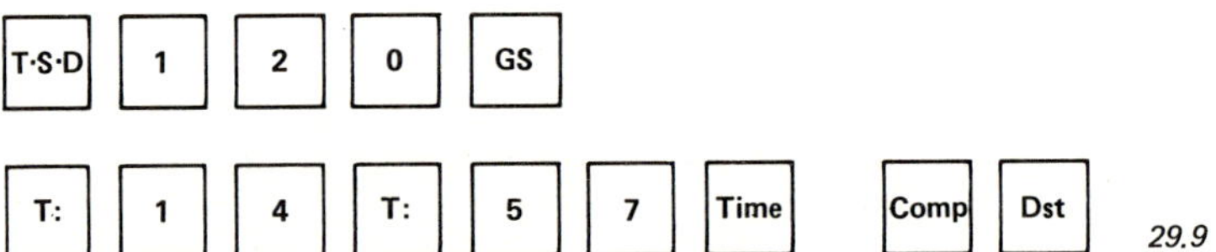

FIGURE 17-3   The Avstar uses the T-S-D program to figure distance to climb.

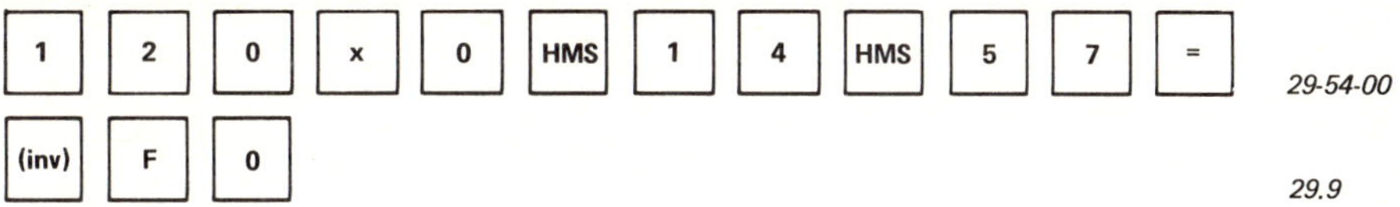

FIGURE 17-4 The N60 multiplies ground speed by time to figure distance to climb.

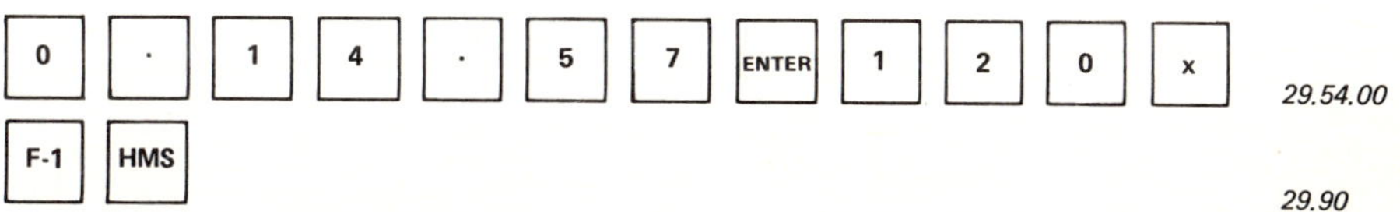

FIGURE 17-5 The OC-1401 also multiplies time by ground speed to arrive at the distance covered during climb.

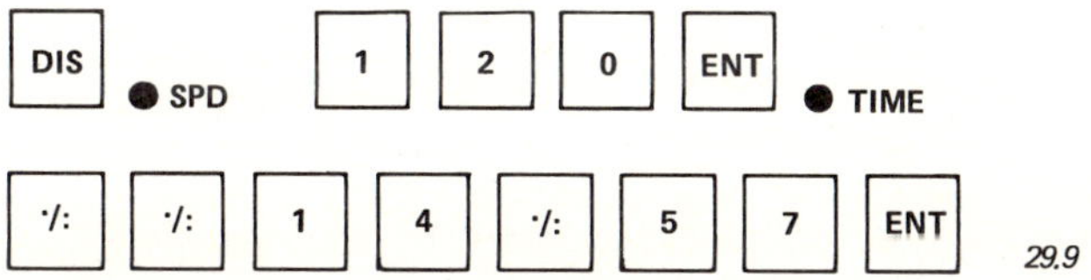

FIGURE 17-6 The Navtronic 1701 uses its distance program to figure distance to climb.

ground speed and its identification second. The result would have been the same. It is for this reason, the fact that the consecutive order in which values are entered into the computer is of no consequence, that the entry of each such value must be followed by pressing an identifying key, as otherwise the computer would have no way of knowing which data stands for which meaning.

With the Commodore N60, in order to figure out the horizontal distance covered in climb the routine is virtually identical to the one used in determining the fuel used (see Figure 17-4). We enter the ground speed (120) and multiply it by the time in hours, minutes, and seconds (0-14-57) and the display shows *29-54-00*. We must now change that to a decimal value by pressing **(inv) F 0**, and the display shows *29.9*, representing the distance covered.

For the OC-1401, if we had wanted to, we could have stored the time figure (0.14.57) in one of the computer's nine memories, in which case we could now simply have recalled it. Failing that, we now

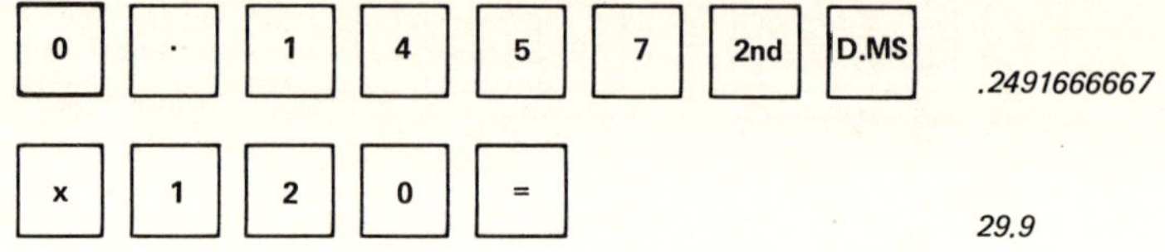

FIGURE 17-7   The TI59 multiplies time by ground speed.

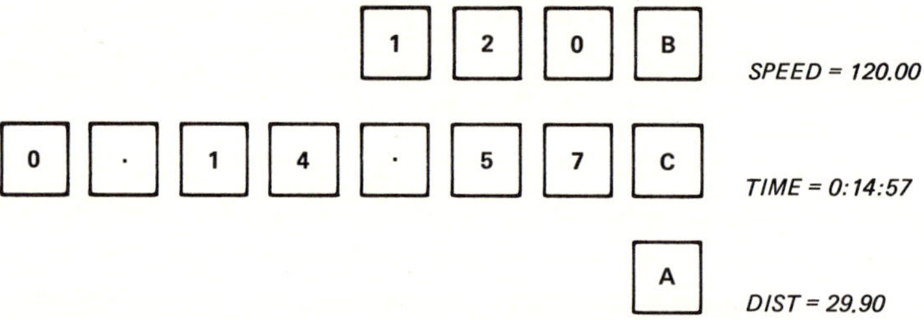

FIGURE 17-8   The HP-41C once more uses the FLIGHT MAN-
AGEMENT program.

enter that figure again into the OC-1401 and then multiply it by the ground speed expected during climb (120) and the result will be displayed as **29.54.00** (see Figure 17-5). Now again pressing **F-1 HMS** will convert that figure to **29.90**, the distance which we can expect to cover while climbing.

When using the Navtronic 1701tr, we first press **DIS** because distance is what we are after (see Figure 17-6). The red light now asks for **SPD** and we enter the ground speed (120) followed by **ENT** which produces a red light next to **TIME**. We then enter the time value (14:57) followed by **ENT** and the display will show **29.9**.

For the TI59 and TI58C, to determine the horizontal distance covered while climbing, we again use the decimal format of the time spent climbing and multiply it by the ground speed and the TI59 (58C) comes up with **29.9** (see Figure 17-7).

The HP-41C still uses the same FLIGHT MANAGEMENT program mode as before. We enter the ground speed (120) followed by the **B** key to identify it as speed (see Figure 17-8). Then we enter the time to climb in hours, minutes, and seconds followed by the **C** key to tell the computer that that figure is a time value. First the display shows **SPEED=120.00**. It then changes to **TIME=0:14:57**. Now, by pressing the **A** key we call up the distance. The display now shows **DIST=29.90**.

18
POINT OF DESCENT

Point of descent is a problem which has taken on increasing importance with the proliferation of turbocharged aircraft and with the continuing increase in the cost of fuel. Turbocharged aircraft, in order to take advantage of the turbocharger's capabilities, will, more often than not, be flown at high altitudes. But it takes quite a bit of time to come down from such high altitudes, especially since too rapid descents tend to cause the engine to cool off to a point where a sudden application of power could result in engine damage, not to mention the fact that the engine might refuse to respond quickly.

Since it would be silly, not to mention wasteful in terms of time as well as fuel, to arrive over our destination airport with 5000 or 6000 feet of altitude still to lose, it is important to determine well in advance where our descent should be initiated in order to get us to pattern altitude by the time we get to our destination, assuming a relatively slow rate of descent and sufficient power setting to keep the engine from cooling off too much.

In the examples illustrated below, we have assumed that we'll be descending from a cruising altitude of 11,500 feet to the pattern altitude of an airport located at an elevation of 87 feet msl. That pattern altitude would be approximately 900 feet msl, meaning that we'd have to descend 10,600 feet. We have further assumed that we'll be descending at 400 ft/min and that we'll maintain the cruise ground speed (150 knots) throughout the descent.

This time, when using the E6b, we can ignore the indicator (see Figure 18-1). We instead place the rate of descent (400 ft/min, shown on the outer rim as **40**) opposite the figure which comes close to the number of feet of altitude which we must lose (**10** for 10,600). Now,

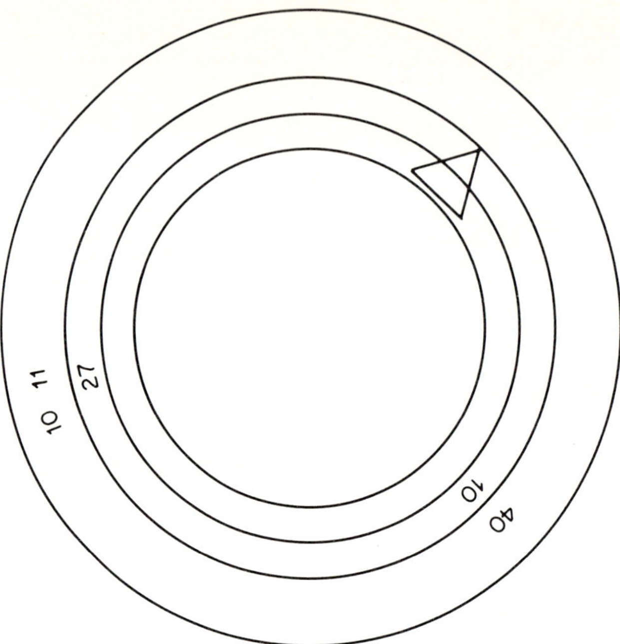

FIGURE 18-1   The first of two calculations performed by
the E6b to determine point of descent.

going around the outer rim of figures we look for another figure representing the 10,600-foot altitude change and we find it between **10** and **11**. Opposite it we'll see the figure **27** which represents the number of minutes we'll be spending in descent.

Now, using the previously described formula, we place the indicator opposite the ground speed (**15** meaning 150 knots) and then, opposite **27** minutes on the second ring of figures, we'll find approximately **67** on the outer rim, meaning that we'll have to start our descent 67 nautical miles from our destination airport (see Figure 18-2).

This can also be accomplished with the standard calculator, using the following steps: We punch in the cruising altitude and deduct from it the pattern altitude at the destination (see Figure 18-3). The display will read *10600*. We now punch in 400 and press the divide (÷) key, producing a display showing *26.5* which represents the time spent during descent in minutes. We now store that in the memory by pressing **MS**. Then we punch in the ground speed (150), press **ENT**

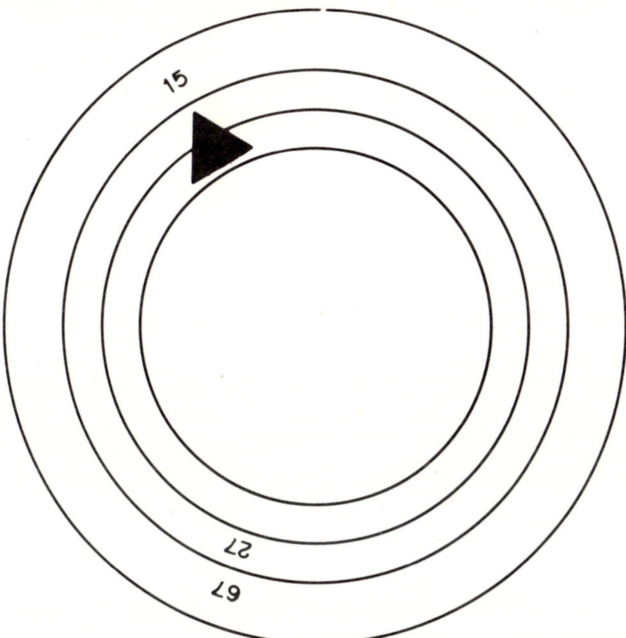

FIGURE 18-2 The second calculation involved in determing point of descent, using the E6b.

followed by 60 and ÷ with the display now showing *2.5* which represents the nautical-miles-per-minute figure. We now recall the stored value from memory by pressing **MR** followed by the multiply key (**x**) and the display now reads *66.25*, representing the nautical miles from our destination where we must start our descent in order to arrive at pattern altitude by the time we get to where we want to go.

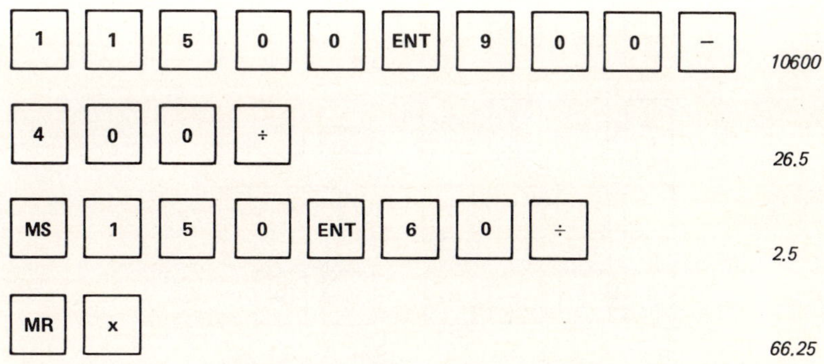

FIGURE 18-3 Using the calculator to determine point of descent.

By using the Avstar the number of steps involved is just about the same, though the routine appears a bit simpler (see Figure 18-4). This being another time-speed-distance problem we start by pressing the **T-S-D** key. We then punch in the cruising altitude and deduct from it the pattern altitude, producing a display showing *10600*. We identify it to the computer as vertical distance by pressing the **Dst** key. We now enter the rate of descent (400 ft/min) followed by **FPH** for identification and follow it by pressing **Comp** and **Time**. The display now shows *26-30-00* representing 26 minutes and 30 seconds. We now enter this time again into the computer, but this time putting minutes and seconds into the correct spaces (if we had used the previously displayed time figure, the computer would have assumed that we were talking about 26 hours and 30 minutes rather than 26 minutes and 30 seconds) and the display will show *0-26-30*. We now punch in **Time** to identify that value and follow it by entering the ground speed (150), identifying it by pressing **GS**, and then follow that by pressing **Comp** and **Dst** which will produce a display showing *66.2*, the distance in nautical miles.

There is also a simpler and shorter way to accomplish the same thing. By the time we are halfway through the routine and the display shows *26-30-00*, we can follow it by pressing **Time** and then punch in the ground speed (150), identify it by pressing **GS**, and then press **Comp** and **Dst** and the display will come up with *3975.0* which is obviously a ridiculous figure. What this figure actually represents is the distance which would have been covered by flying 26 hours and 30 minutes at a ground speed of 150 knots. In order to get the right answer we now divide that figure by 60, and the resultant display will

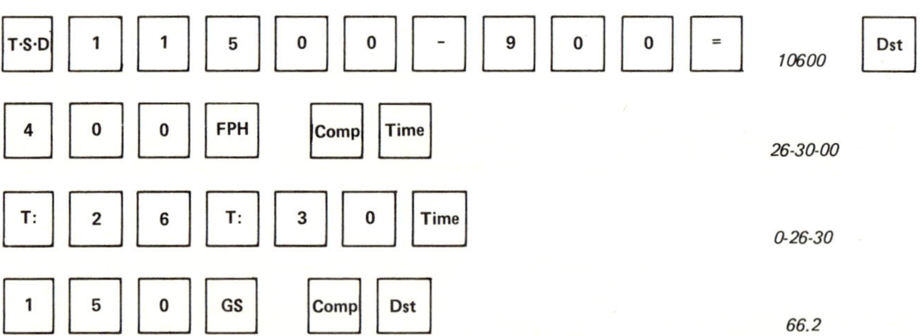

FIGURE 18-4   The Avstar employs the T-S-D program along with calculator functions to figure point of descent.

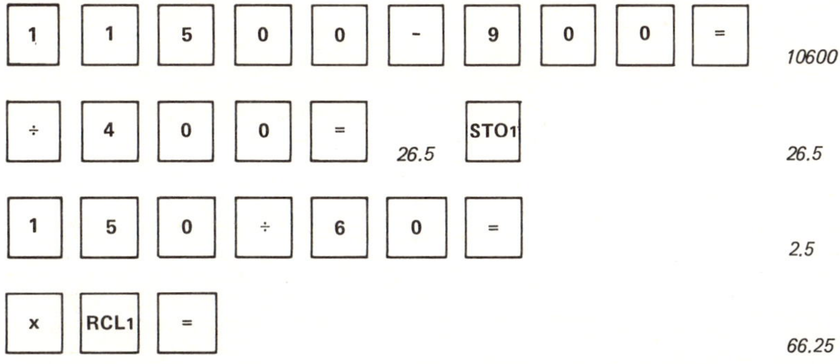

FIGURE 18-5 The N60 makes use of one of its memories in figuring point of descent.

show **66.25**, which is just about the same as what we arrived at the other way.

The Commodore N60 has no program to come up with the point of descent. As a result it is strictly a mathematical problem (see Figure 18-5). To arrive at the solution we enter the cruise altitude (11,500) followed by − and the pattern altitude at the destination airport (900) followed by = and the display gives us the amount of altitude change: **10600**. We divide that by the rate of descent (400), resulting in a display showing **26.5**. We store that figure in the memory by pressing **STO**$_1$. We then enter the ground speed (150) and divide it by 60 in order to arrive at the ground speed in terms of miles per minute. The display now shows **2.5**. We now press the multiply key followed by the **RCL**$_1$ key which commands the computer to multiply the last display with the value stored in the computer and by pressing = we see the display changing to **66.25**, the distance before reaching our destination at which point the descent should be initiated.

When using the OC-1401 we start by deducting the pattern altitude of the destination airport (900) from the flight altitude (11,500) and divide that result (10,600) by the rate of descent (400) and the display comes up with 26^1/$_2$ minutes in decimal form, **26.5**, as the time required for the descent (see Figure 18-6). We now multiply that by the expected ground speed (150) and divide the resulting display (**3975.00**) by 60, since we are now dealing with hours rather than minutes, and the display shows that we have to start our descent **66.25** nautical miles from the destination airport.

With the Navtronic 1701tr the routine is quite a bit simpler (see

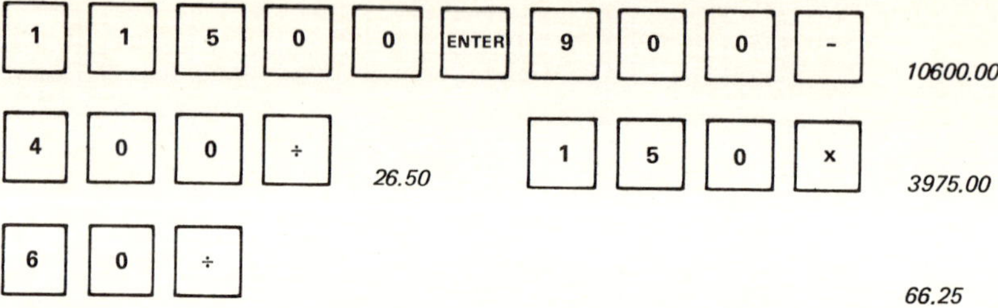

FIGURE 18-6   Using the OC-1401 in figuring point of descent.

Figure 18-7). We start by pressing **ROC** (rate of climb) to identify the problem as concerning vertical as well as horizontal distance. The red light now asks for **DIST** which, in this case, refers to the rate of descent and we punch in 400 followed by **ENT**. This produces a red light next to **SPD** and we punch in the ground speed (150) followed by **ENT**. Now the red light appears next to **ALT CHG** (altitude change) and we punch in 11,500 minus 600 which produces a display showing *10600*. We now press **ENT** and the display shows *66.3* as the distance in nautical miles from our destination where the descent must be initiated.

Calculating point of descent with the TI59 (58C) is best accomplished by using the TI59 (58C) strictly as a calculator. We deduct the pattern altitude at the destination airport from the cruising altitude and divide the result by the rate of descent (see Figure 18-8). The result appears as minutes and decimals of minutes. We now multiply that by the ground speed and then, since ground speed is figured in hours and the rate of descent in feet per minute, we divide the result by 60 (the number of minutes in an hour) and the display shows *66.25* as the distance from our destination where we should set up a 400 ft/min rate of descent, retaining the 150-knot ground speed.

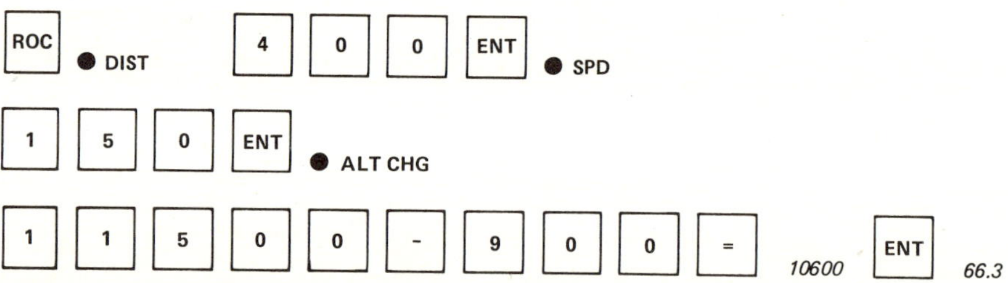

FIGURE 18-7   The Navtronic 1701 uses its rate-of-climb program as part of the routine of figuring point of descent.

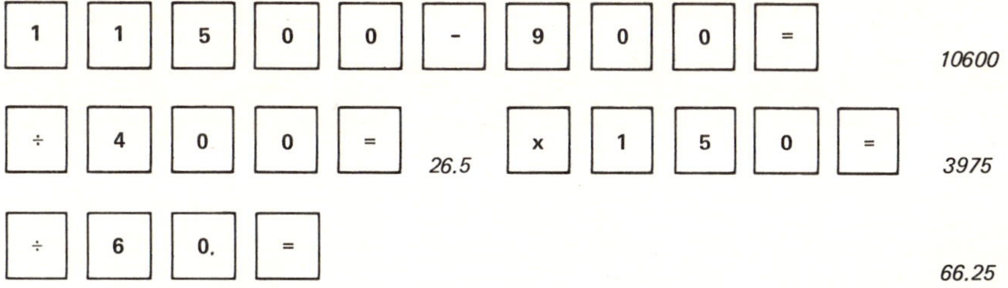

FIGURE 18-8   With the TI59 point of descent is arrived at by simple calculations.

With the HP-41C we still use the same FLIGHT MANAGEMENT program mode we've been using all along. We enter the cruise altitude and subtract from it the pattern altitude at the destination, and the resultant display shows *10600.00* (see Figure 18-9). By pressing **A** we identify that as the vertical distance. We now enter the rate of descent and press **B** to identify it as the vertical speed. Pressing **C** will now produce the time spent in descent in terms of minutes, seconds, and sixtieths of a second: *26 : 30 : 00*. We now correct that figure to hours, minutes, and seconds by entering **0 . 26 . 30** followed by the **C** key.

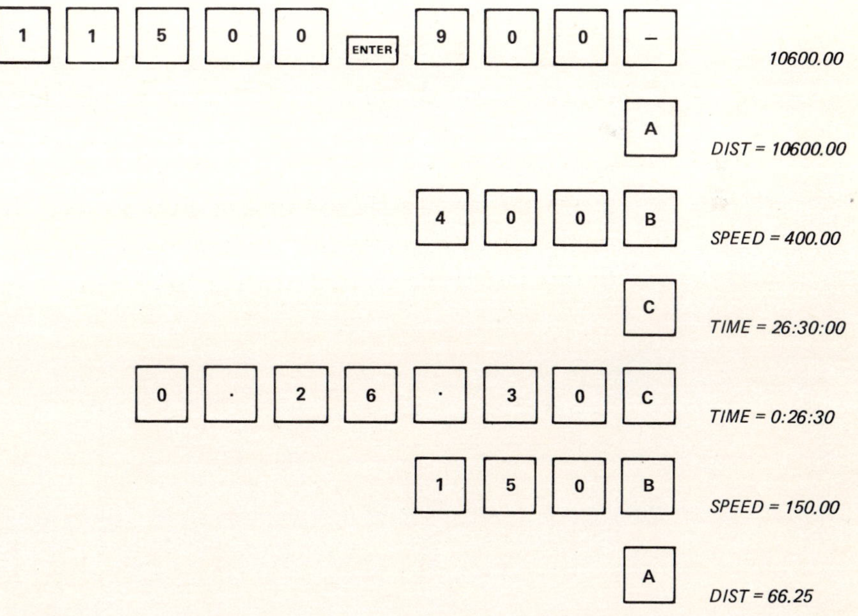

FIGURE 18-9   The HP-41C uses the FLIGHT MANAGEMENT program to calculate point of descent.

Then we enter the ground speed and identify it by pressing **B**. Now we press **A** and the horizontal distance from the destination at which descent should be initiated is displayed as *DIST=66.25*.

Again, as was the case with the Avstar, there is an alternate way of achieving the same result. Once we have reached the point where the display shows *TIME=26:30:00* we can continue by entering the ground speed (150) and press **B** and then we press **A** and the display will show *DIST=3975.00*, which represents the distance which would have been covered in 26 hours and 30 minutes. We now press 60 and the divide key (÷) and the display changes to the correct **66.25** figure, this time without being preceded by the identifier *DIST=*.

19
FUEL USED IN DESCENT

The fuel which we use during the descent, though we like to think of it as insignificant, does have to be figured in if we want to have a realistic picture of the overall fuel consumption from takeoff to touchdown. In the following examples I have assumed a fuel flow at the reduced power setting of 12 gal/h. And we, of course, know already from the preceding computation that the descent will take us 26.5 minutes. By simply taking a rough guess we'll figure that it will take a bit under 6 gallons.

When using the E6b we place the indicator opposite the fuel flow (**12**) and then we find the time spent in descent on the second ring (see Figure 19-1). This will be halfway between **26** and **27**, and opposite it we find the figure **53**. Well, quite obviously we'll not be using 53 gallons, so the answer is that in this instance the 53 stands for 5.3.

With the ordinary calculator we punch in the fuel flow (12) and, because it is in hours, but the time value we'll be working with is in minutes, we'll now divide that figure by 60 and the display will show *.2* (see Figure 19-2). We now multiply that by the time (26.5) and the result will show *5.3*, the amount of fuel used.

With the Avstar we could, of course, do exactly the same that we did with the calculator by simply using the calculator portion of the computer. Or we can identify the problem to the computer by pressing the **T-S-D** key and then entering the time (26:30) and the display will show *0-26-30* (see Figure 19-3). We identify that value by pressing **Time**, then punch in the fuel flow (12), identify it by pressing **FPH** followed by **Comp** and **Fuel** and the display will again show *5.3*. The advantage in using the Avstar as a computer rather than as a simple calculator in this case is that we can use hours:minutes:seconds figures instead of having to express time in terms of decimals.

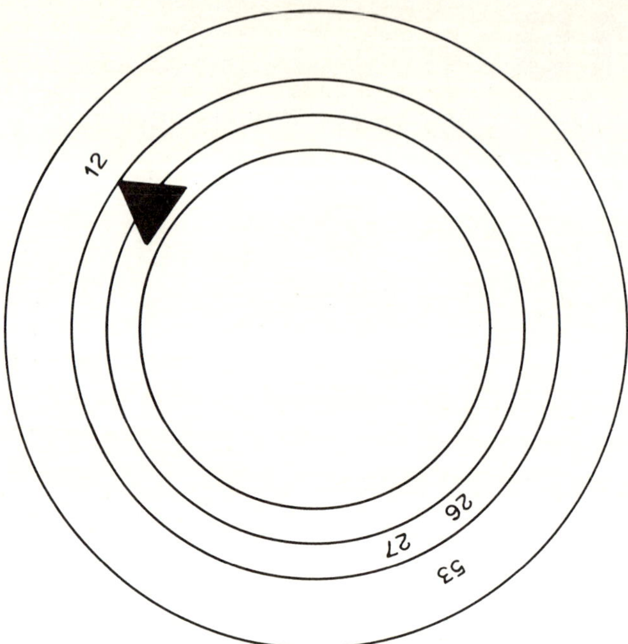

FIGURE 19-1   Fuel used in descent, figured with the E6b.

When using the Commodore N60 we multiply the time by the fuel flow and then convert the resulting figure into decimals, just as was done when determining the amount of fuel used during climb (see Figure 19-4).

With the OC-1401, remembering that it takes 26.5 minutes to descend from flight level to the pattern altitude at the destination airport, we must now multiply that figure by the fuel flow expected during descent (12) and then we again have to divide the result (318.00) by 60 in order to allow for the fact that fuel flow is expressed in gallons per hour, while the initial value was expressed in minutes

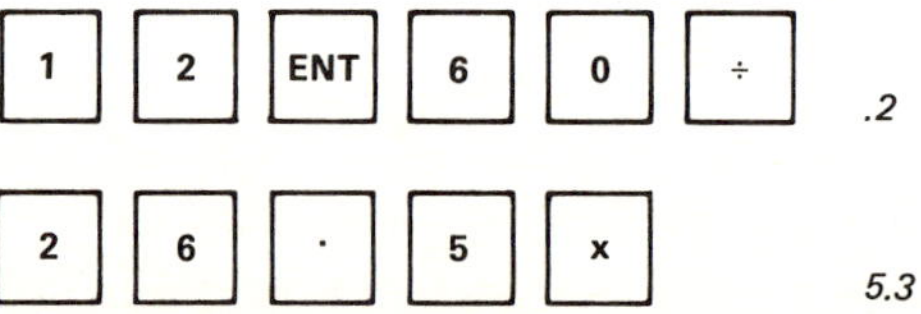

FIGURE 19-2   Fuel used in descent, figured with the calculator.

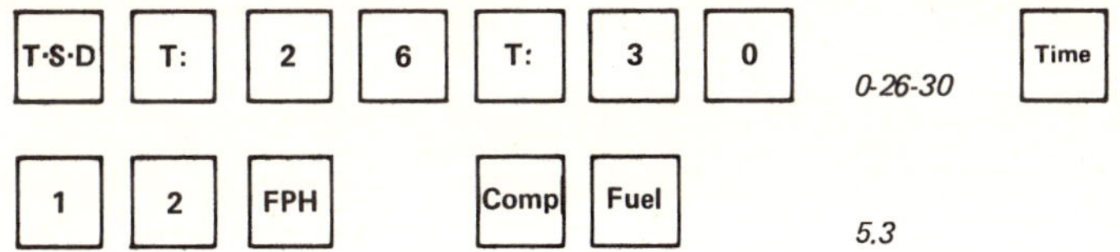

FIGURE 19-3   The Avstar again uses the T-S-D program to figure fuel needed during descent.

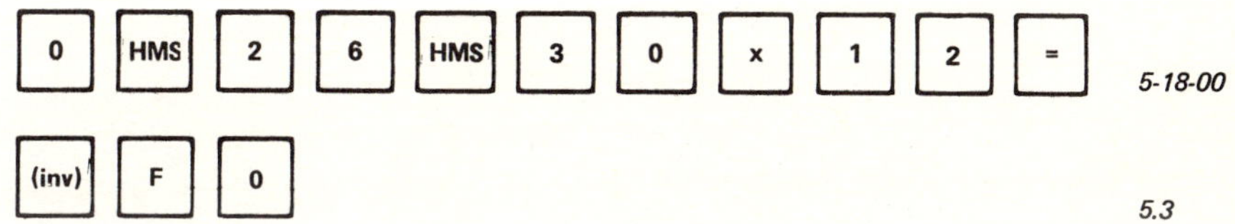

FIGURE 19-4   The N60 multiplies time by fuel flow.

(see Figure 19-5). The resulting display then is *5.30*, the amount of fuel burned during descent.

With the Navtronic 1701tr we start by pressing **FUEL** which activates the red light next to **RATE**. We punch in the fuel flow (12) and pressing **ENT** will cause the red light next to **TIME** to come on. We now enter the time (26:30) and pressing **ENT** will produce *5.3* in the display window (see Figure 19-6).

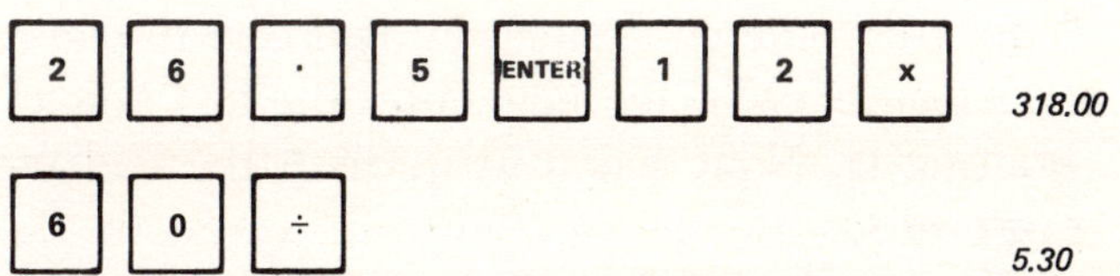

FIGURE 19-5   The OC-1401 also multiplies time by fuel flow.

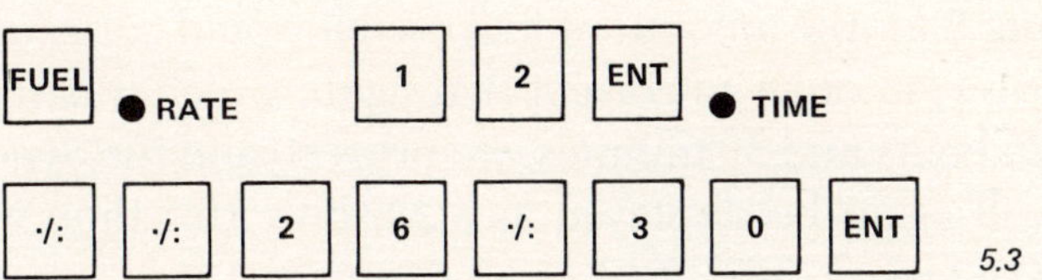

FIGURE 19-6   The Navtronic 1701 uses its fuel program to figure fuel used in descent.

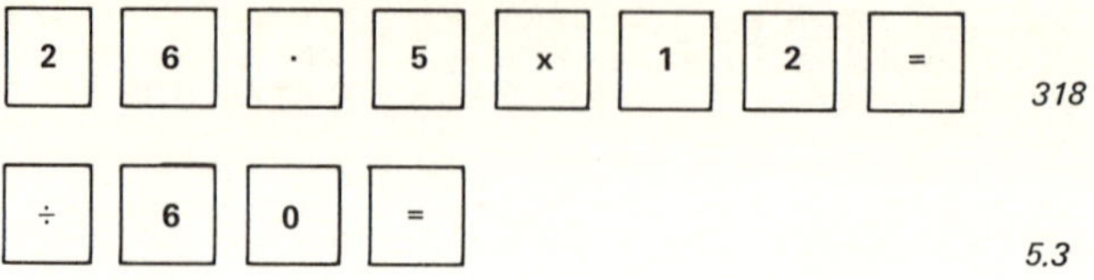

318

5.3

FIGURE 19-7   The TI59 uses simple calculation to figure fuel used in descent.

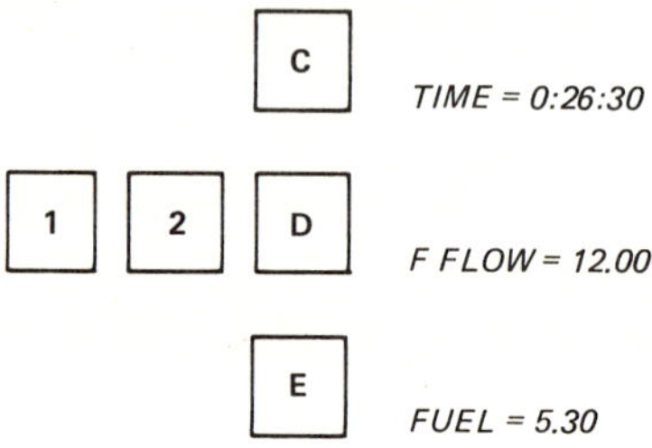

FIGURE 19-8   With the HP-41C the problem is solved once more by the FUEL MANAGEMENT program.

With the TI59, fuel used in descent is figured exactly the same way as fuel used in climb: Time, in decimal format times fuel flow, divided by 60 (see Figure 19-7).

With the HP-41C we'll find that when we press **C** for time, the previously entered figure will again be displayed as *TIME=0:26:30* (see Figure 19-8). We now enter the fuel flow (12) and identify it by pressing **D**. Now, when we press **E** the fuel used will be shown as *FUEL=5.30*.

If we had used the alternate method of determining the point of descent, as was described in the last chapter, then, when pressing **C**, the display would have shown *TIME=26:30:00*. In that case we can again follow that by entering the fuel flow (12) and then, when pressing **E** for the amount of fuel used it would come up with *FUEL=318.00*. Now, in order to correct that figure which is based on a time factor of 26 hours and 30 minutes, we press 60 and the divide key and the result will again be displayed as *5.30*, only this time without the identifier *FUEL=*.

20

DISTANCE, TIME, AND FUEL FROM TAKEOFF TO LANDING

So far we have broken up the flight into its individual portions. But what we really want to know is the total amount of time required to cover the entire distance including climb, cruise, and descent, and the total amount of fuel which we'll be using. Simply adding the results of the previous computations would result in an erroneous figure, because we never did figure out how much time and fuel we'd be expending during the cruise portion of the flight. This would require deducting the distances covered in climb and descent from the total distance and then figuring the time and fuel for the remaining distance and adding the results to those obtained when we worked out the climb and descent problems. But there are a number of simpler ways to arrive at the desired result.

While it is possible to obtain the desired information by using the E6b, the number of steps involved is so numerous and the results so imprecise, that I sincerely doubt that anyone would want to bother.

Using a regular calculator to solve this problem, while possible, involves two separate and rather lengthy calculations (see Figure 20-1). But just to illustrate how it can be done, we shall run through all of the steps, using the figures which we had obtained earlier when using the calculator to solve the previous problems.

We start by punching in the total distance (245) and deducting the distance covered in climb (29.90666) which produces a display showing *215.09334*. We now deduct the distance covered during the descent (66.25) and the display changes to *148.84334* which is the distance in nautical miles which will be covered while we are at cruise altitude. We now divide that figure by our ground speed (150) and the display will show *.99228893*. That figure is the amount of time spent in

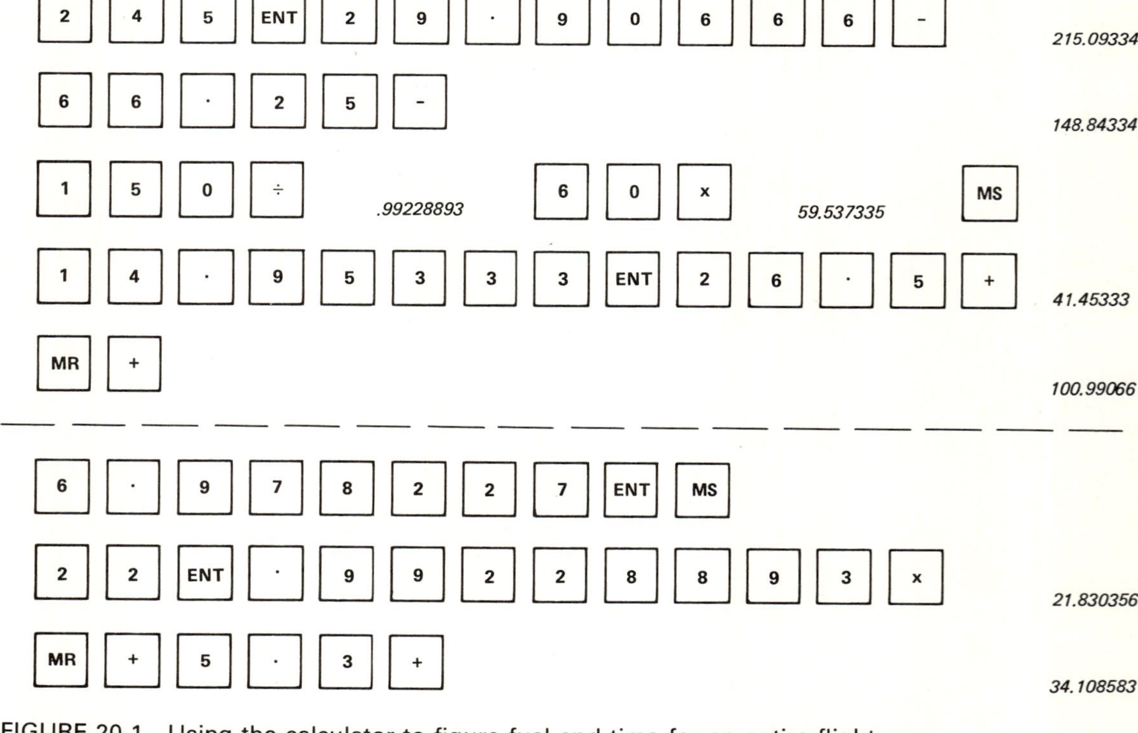

FIGURE 20-1   Using the calculator to figure fuel and time for an entire flight.

cruise, expressed in decimals of an hour. To convert it into minutes we multiply that figure by 60 and the display shows *59.537335*, meaning that we'll spend about 59½ minutes in cruise. We now punch the **MS** key to store that figure in the memory and we then enter the time spent climbing (14.95333 minutes) and add to it the time spent in descent (26.5 minutes). We then press the **MR** key to retrieve the value stored in the memory and by adding it to the above the display will show *100.99066* which means that the entire trip from takeoff to destination will take approximately 101 minutes or 1 hour and 41 minutes.

The second calculation deals with the fuel used. Here we start off by entering the amount of fuel burned in climb (6.978227). We now store that figure in the memory by pressing **MS** and then we enter the fuel flow at cruise-power setting (22 gal/h) and multiply it by the time spent in cruise, expressed in decimals of the hour (.99228893, the figure which we had obtained in the previous calculation) which will result in a display showing *21.830356*, representing the fuel burned during the cruise portion of the flight. We now retrieve the previous figure from the memory by pressing **MR**, add it to the above and then add the fuel used during descent (5.3) and the display will show *34.108583*, meaning that we'll be using a bit over 34 gallons for the entire trip.

As can be seen by the above, using a calculator for problems which involve multiple calculations can become rather complicated, the biggest problem being that we have to remember all those steps, and the consecutive order in which they must be executed.

Things get a bit simpler when we use the Avstar for this purpose. Again, in these illustrations, we'll use the results exactly as they were obtained by the Avstar during the previous computations (see Figure 20-2).

We start by pressing **T-S-D**. We then enter the total distance (245) and deduct from it the distances covered during climb and descent (29.9 and 66.2) and the display shows the distance covered in cruise as *148.9* nautical miles. We identify that value to the computer by pressing **Dst** and then enter the ground speed (150) and identify it by pressing **GS**. We now press **Comp** and **Time** and the display will show that the time in cruise will be *0-59-34*. We now press **Sto** in order to store that value in the memory while we perform the next computation. We press **Time**, then enter the fuel flow (22 gal/h) and identify that figure by pressing **FPH** and follow that by using the **Comp** and **Fuel**

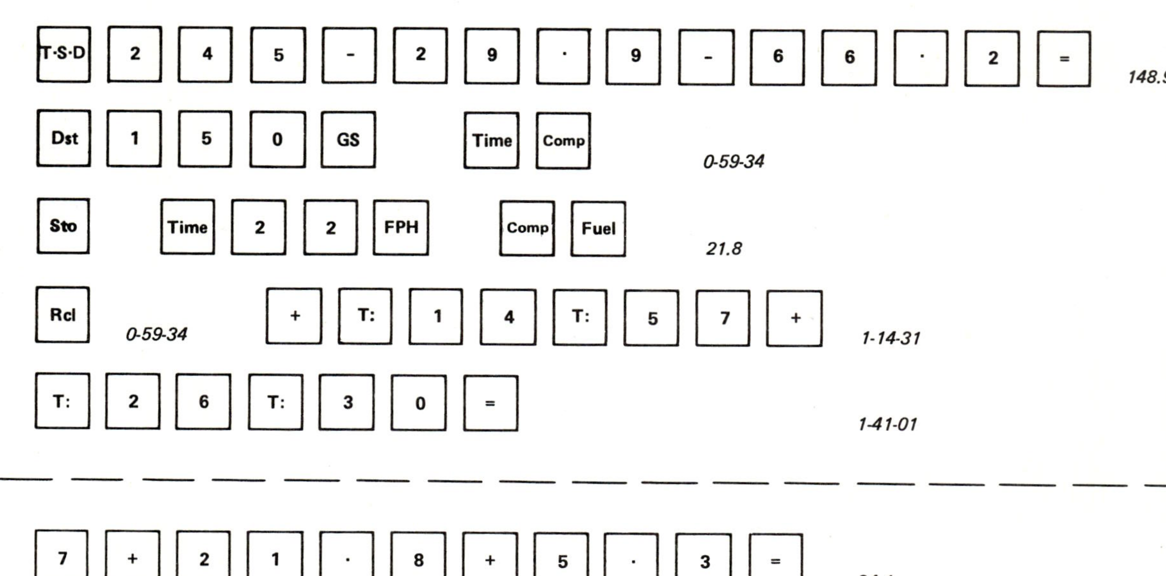

FIGURE 20-2   Using the Avstar to figure time and fuel from takeoff to landing.

keys. The display now shows *21.8* representing the amount of fuel used during the cruise portion of the flight. We now recall the previously stored value from the memory by pressing **Rcl** and the display once more shows *0-59-34*. We add to it the time spent during climb and descent (14:57 and 26:30) and the resultant display shows *1-41-01* as the total time for the entire flight.

To figure the total amount of fuel used, we now simply add the fuel used in climb (7) to the fuel used during cruise (21.8) and then add the fuel used during descent (5.3) and the display will show *34.1* as the total for the fuel used.

There is not much difference in the number of keystrokes involved, when using the Commodore N60, but, like with the Avstar, the availability of a second memory simplifies the task somewhat. We deduct the horizontal distances covered during climb and descent from the total distance and find that the distance at cruise altitude will be *148.85* (see Figure 20-3). We divide that by the ground speed and the result appears as *0.992333333*. By using the **F** and **0** keys we convert that into *0-59-32* and that we now store in the memory. As soon as the value is stored by using the **STO₁** key, the display will automatically revert to *0.992333333*. We convert that a second time to *0-59-32* and add to that figure the time spent during climb and descent, arriving at a total time en route of *1-40-59*.

Now, in order to get the total amount of fuel burned, we recall the value stored in the memory by pressing **RCL₁** and multiply it by the cruise fuel flow which results in a display of *21.80933333*. To this we add the fuel used during climb and descent and the final figure for the amount needed for the whole trip is displayed as *34.08933333*.

With distance, time, and fuel calculations the multiple memory capability of the OC-1401 comes in handy. We start by deducting the horizontal distances covered during climb and descent from the total distance which produces a display of *148.85*, the distance we'll be covering at cruise altitude (see Figure 20-4). By dividing that by the expected ground speed (150) we produce a display of *0.99*. We now press **F H** and the display changes to *0.59.32*, representing the time spent at cruise level. By pressing **F STO 1** we place that figure into memory number 1. We now punch in the time spent in climb (0.14.57) and add to it the time spent in descent (0.26.30) and the display now shows *1.40.59*. It must be remembered that even though we entered the time spent at cruise into the memory, it also remained

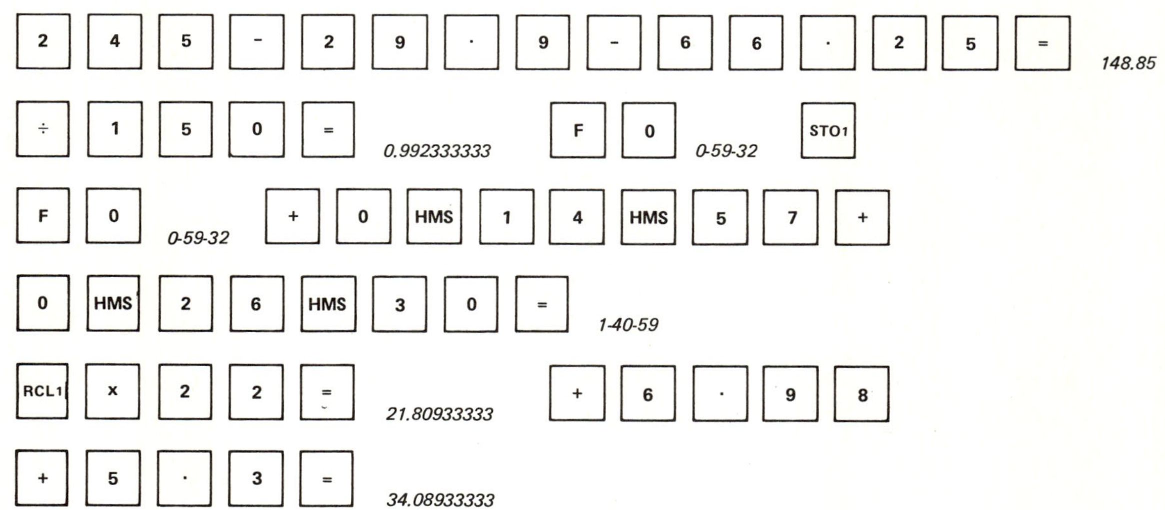

FIGURE 20-3   The N60 comes up with a large number of decimals if figuring time and fuel for an entire trip.

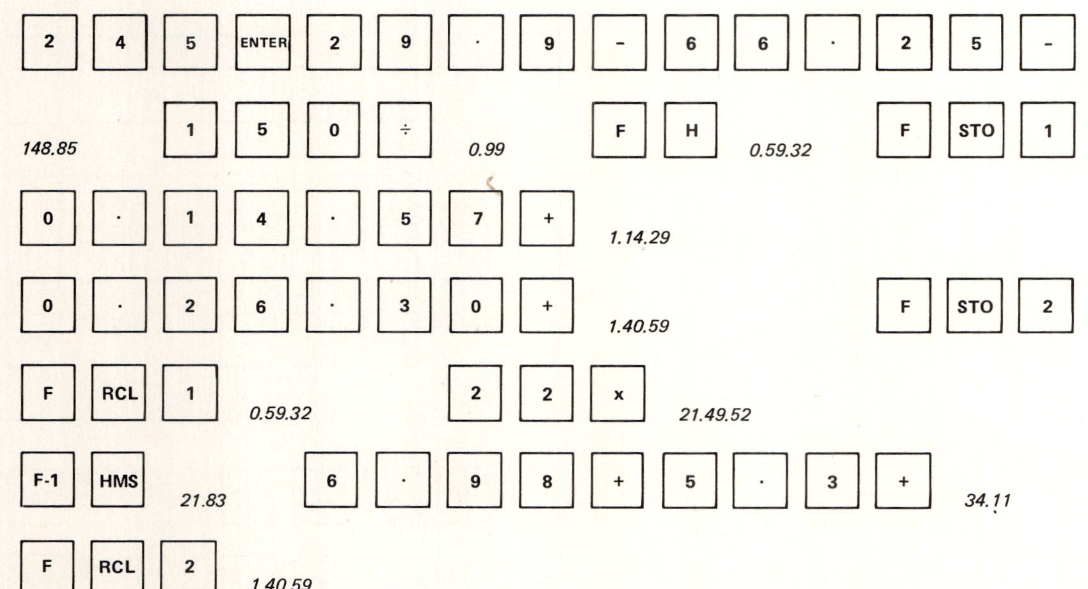

FIGURE 20-4   With the OC-1401 its multiple memories come in handy when figuring time and fuel for a trip.

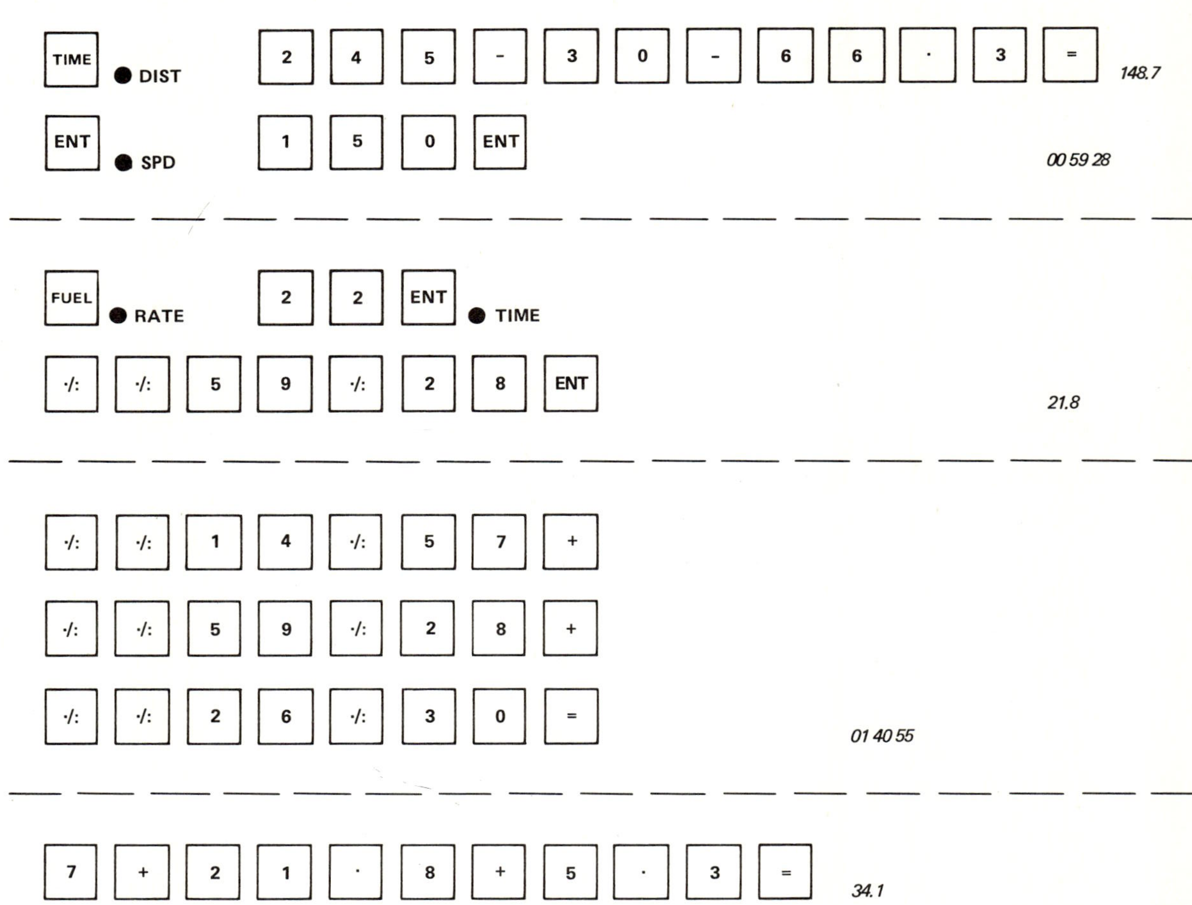

FIGURE 20-5   With the Navtronic 1701 it takes four separate calculations to arrive at time and fuel figures for an entire trip.

in the display, and what we have actually done is to add to it the climb and descent times. Thus, the current display (*1.40.59*) is the total time for the entire trip. We may now store that figure in one of the memories for future recall, by pressing **F STO 2**. We now recall the value stored in memory earlier by pressing **F RCL 1** and when it appears in the display (*0.59.32*) we multiply it by the fuel flow in cruise (22) which produces a display of *21.49.52*. We convert that figure by pressing **F-1 HMS** and the new display tells us that we'll be using *21.83* gallons during the cruise portion of the flight. To this we add the fuel used during climb and descent (6.98 + 5.3) and the new display represents the total amount of fuel burned during the entire flight. I might point out here that the totals for fuel and time could have been placed into memory positions after completing each of the previous preflight computations. In that case, instead of having had to enter those values from memory or from notes made with pencil and paper, we could simply have recalled them from their appropriate memory positions. Now, knowing that it will take 34.11 gallons of fuel to make that flight, we might once more want to look at the total time. It is still stored in memory number 2 and can be recalled by pressing **F RCL 2**.

Since the Navtronic 1701tr does not have a memory capability, arriving at the various totals calls for four different computations (see Figure 20-5). First we want to know how much time we'll be spending in cruise. We press **TIME** and the red light asks for **DIST**. We punch in the total distance (245) and deduct from it the climb and descent distances (30 and 66.3) and the display now shows *148.7*. We enter that by pressing **ENT** and the red light asks for **SPD**. We put in the ground speed (150) followed by **ENT** and the display shows *00 59 28*.

Next we want to know the fuel burned in cruise. We press **FUEL** and the red light asks for **RATE**. We put in the fuel flow (22) and when we press **ENT** the red light asks for **TIME**. We enter the time (59:28) and the resultant display shows *21.8*, representing the fuel burned in cruise.

We now want to know the total time of the flight. We simply add the three time values in minutes and seconds to one another (14:57 plus 59:28 plus 26:30) and the display comes up with the total time: *01 40 55*.

Now, to obtain the fuel total, we again simply add all of the fuel figures (7 plus 21.8 plus 5.3) and the display shows the result: *34.1* gallons.

The problem here is that we have to have pencil and paper handy in order to remember the result of one computation while we perform the next ones.

I have not included the TI59 and the HP41C in this chapter. Both are, of course, capable of providing this information, but both have special programs for this purpose which have been discussed in detail in Chapters 9, 10, and 14.

21
ENDURANCE

Here we want to know how long we can stay in the air at a given rate of fuel consumption before the tanks run dry. The problem, of course, is a simple one. All that is called for is to divide the amount of fuel on board by the known fuel flow in terms of gallons (or pounds) per hour.

In our examples I have assumed that the airplane carries 84 gallons of fuel, and that the average fuel flow is 13 gal/h.

With the E6b we place the indicator opposite the fuel flow (**13**) and then, opposite the total fuel on board (**84**, or just next to **85**) we can read the endurance on the inner scale which here seems to point to approximately $6^{1}/_{2}$ hours (see Figure 21-1).

Using the standard calculator we simply divide the fuel on board (84) by the fuel flow (13) and the result will be displayed as *6.4615384* hours (see Figure 21-2). If we now want to know what those decimals of an hour represent in terms of minutes, we deduct the hours (minus 6) and multiply by 60, and the display will show *27.692304* minutes, meaning that the total endurance is 6 hours and 28 minutes, or just a bit less.

When using the Avstar, we identify the program by pressing **T-S-D**, then enter the fuel on board (84) and identify it by using **Fuel**, then add the fuel flow (13) followed by the identification **FPH** (see Figure 21-3). Then using **Comp** and **Time** the display will come up with *6-27-42*.

When using the N60 we again simply divide the total amount of fuel by the fuel flow and convert the resulting display, *6.461538462* into hours, minutes, and seconds by pressing **F** and **0** (see Figure 21-4). The time then displayed is *6-27-42*.

With the OC-1401, endurance is the result of dividing the available

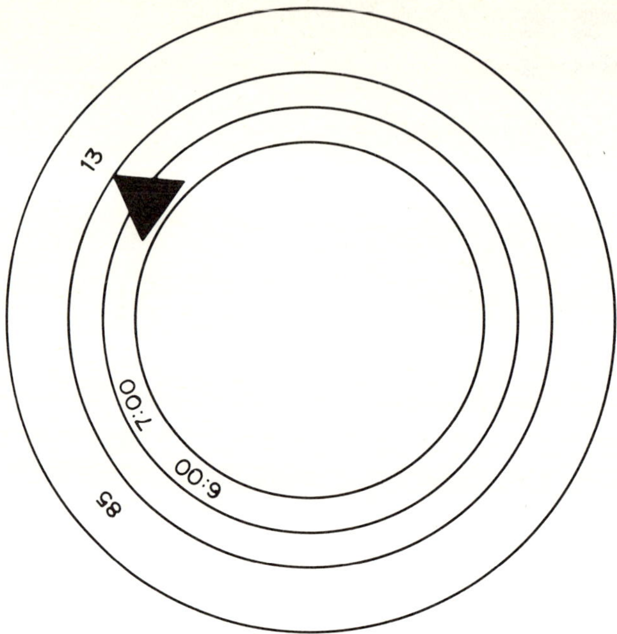

FIGURE 21-1   Endurance figured with the E6b.

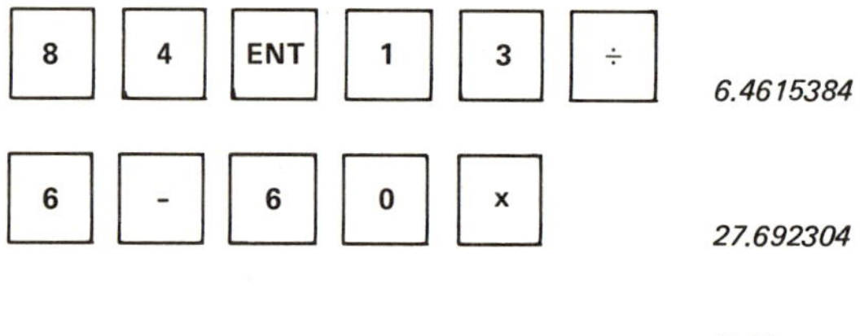

FIGURE 21-2   Endurance arrived at by calculator.

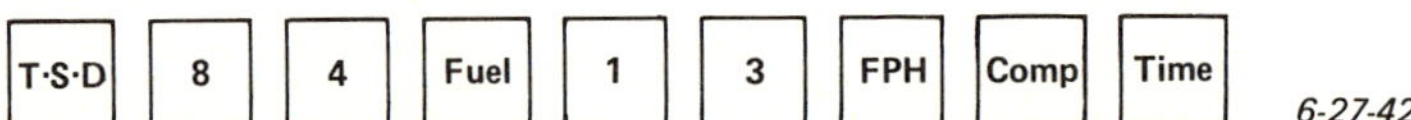

FIGURE 21-3   Endurance figured by using the T-S-D program in the Avstar.

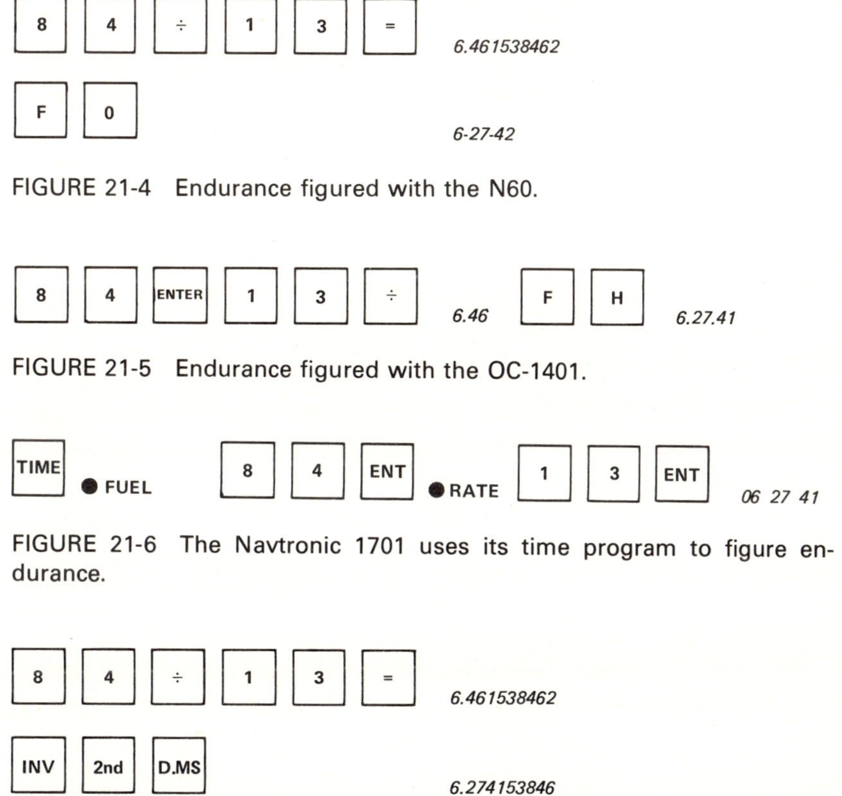

FIGURE 21-4   Endurance figured with the N60.

FIGURE 21-5   Endurance figured with the OC-1401.

FIGURE 21-6   The Navtronic 1701 uses its time program to figure endurance.

FIGURE 21-7   Endurance figured by using the TI59.

fuel (84) by the fuel flow (13) and then converting the resulting display (*6.46*) to an hours:minutes:seconds format by pressing **F H** (see Figure 21-5). The endurance at that rate of fuel consumption is *6.27.41* or 6 hours, 27 minutes, and 41 seconds. If the next computation deals with the range, based on the above figures, we might store that last figure in one of the memories. (Remember, data stored in the memories remains there only as long as the computer is not turned off.)

With the Navtronic 1701tr we start by pressing **TIME** which results in the request **FUEL** (see Figure 21-6). We enter 84 and are now asked to provide **RATE**. We enter the fuel flow (13) and the display shows *06 27 41*.

With the TI59 (58C), again a simple calculation brings the desired result: Fuel quantity divided by fuel flow results in time in decimals (see Figure 21-7). When that figure, *6.461538462*, is converted by the

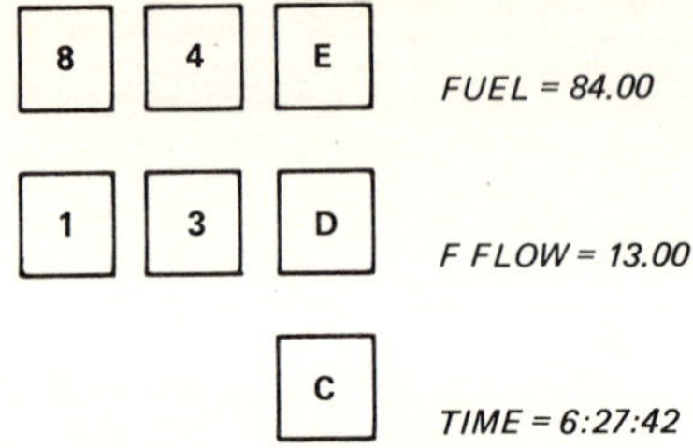

FIGURE 21-8 The FLIGHT MANAGEMENT program in the HP-41C is used to figure endurance.

TI59 (58C) the display changes to *6.274153846* which, for all practical purposes is 6 hours, 27 minutes, and 42 seconds.

The HP-41C, still in the FLIGHT MANAGEMENT program mode, requires only that we enter 84 and identify that as the amount of fuel by pressing **E**, and then enter the fuel flow (13) and identify it by pressing **D** (see Figure 21-8). Now, pressing **C** commands the computer to come up with a result in terms of time, and the display will show *TIME=6:27:42*. Simple.

Remember that the HP-41C never forgets. So if you want to know the endurance assuming different rates of fuel flow, all that is needed is to change the fuel-flow figure and the computer will come up with new results. Similarly, if you want to know how much fuel you need in order to fly a certain length of time at the previously entered fuel flow, just change the time figure and then call up the amount needed by pressing **F**.

22
RANGE

How far can we go assuming a certain average ground speed? Again the problem is simple. We simply multiply the endurance figure arrived at in the previous problem by the ground speed. That's all. Just remember that if the ground speed is in miles per hour, the result will be in statute miles, but if the ground speed is in knots, then the result will be in nautical miles.

Using the E6b we place the indicator opposite the ground speed (**15** here meaning 150) and opposite the endurance (halfway between **6:00** and **7:00**) we find the figure **95** which in this instance represents 950 nautical or statute miles (see Figure 22-1).

Using a standard calculator we first have to transpose 6 hours and 28 minutes into a decimal figure. We enter the minutes (28) and divide them by 60, producing a display of *.46666666* (see Figure 22-2). To this we now add the hours (6) and the display shows *6.4666666* which we now store in the memory. We then enter 150 (the ground speed) after which we recall the stored figure from memory and multiply the two, producing a range figure of *969.99999* miles.

With the Avstar we start with **T-S-D** followed by the ground speed (150) and its identification (**GS**) (see Figure 22-3). We then enter the time in hours, minutes, and seconds (6:27:42), press **Time** for identification, and then use the **Comp** and **Dst** keys to produce the result: *969.2* miles.

To determine the range when using the N60 we multiply the speed (150) by the endurance (6:27:42) and the display will show *969-15-00* (see Figure 22-4). We now convert that meaningless figure to decimals by pressing **(inv) F 0** and the range will appear as *969.25*.

With the OC-1401, we now either enter the endurance in hours:

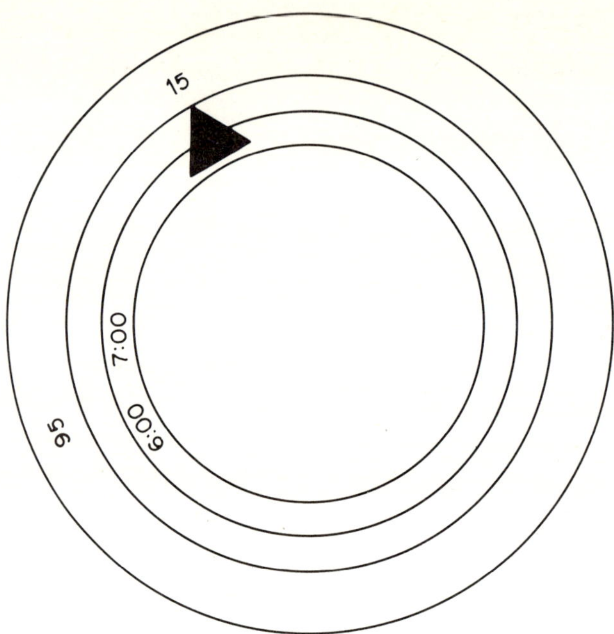

FIGURE 22-1   Range figured with the E6b.

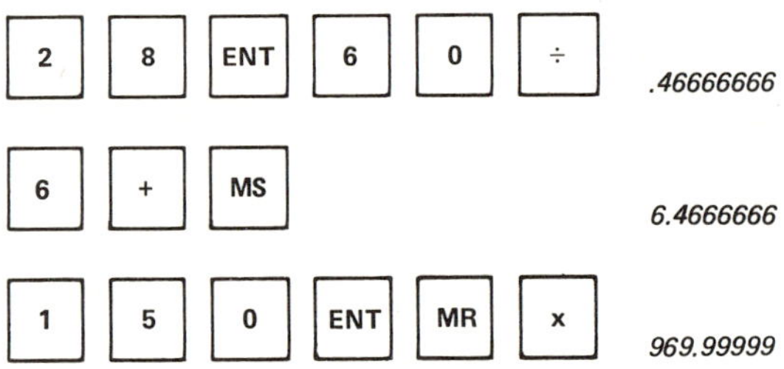

FIGURE 22-2   Range figured by using the calculator.

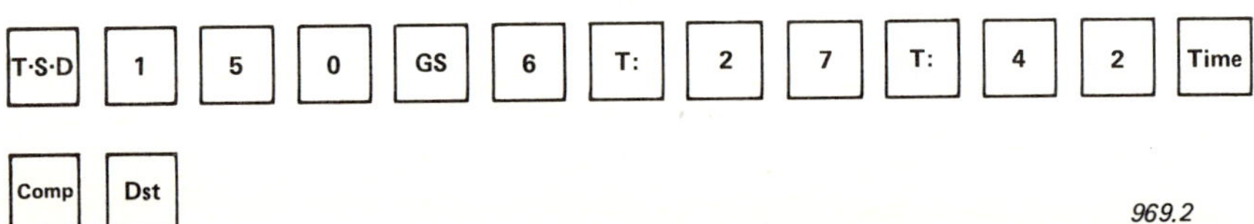

FIGURE 22-3   Range figured by using the T-S-D program in the Avstar.

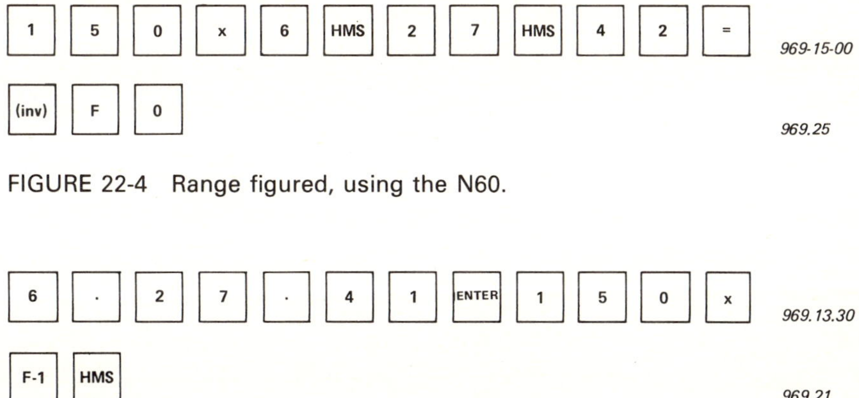

FIGURE 22-4   Range figured, using the N60.

FIGURE 22-5   Range figured, using the OC-1401.

minutes:seconds or we recall that figure from the memory (see Figure 22-5). We then multiply it by the true airspeed (or ground speed, if available) and the result comes up as *969.13.30*. By pressing **F-1 HMS** we now convert that to nautical miles: *969.21*, representing the no-reserve range.

The Navtronic 1701tr requires that we first press **DIS** which activates the red light next to **SPD**. We enter the ground speed (150) and are now asked for **TIME** (see Figure 22-6). We enter the endurance in hours, minutes, and seconds (6:27:41) and the display shows *969.2*.

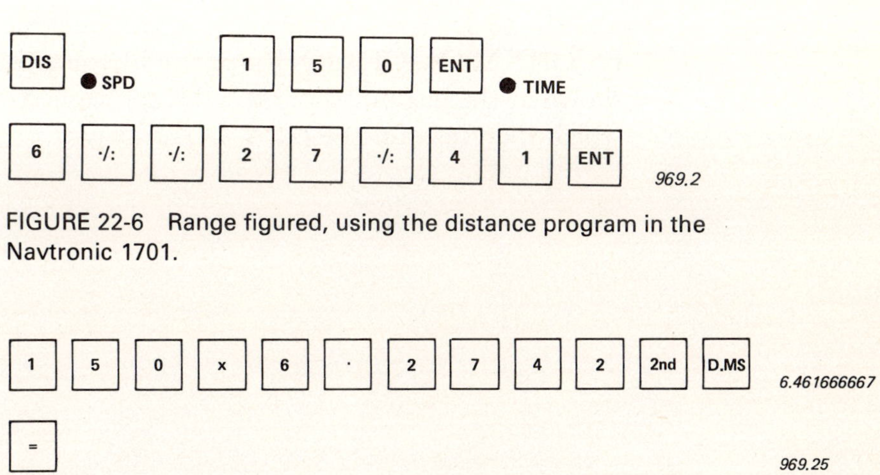

FIGURE 22-6   Range figured, using the distance program in the Navtronic 1701.

FIGURE 22-7   Range figured, using the TI59.

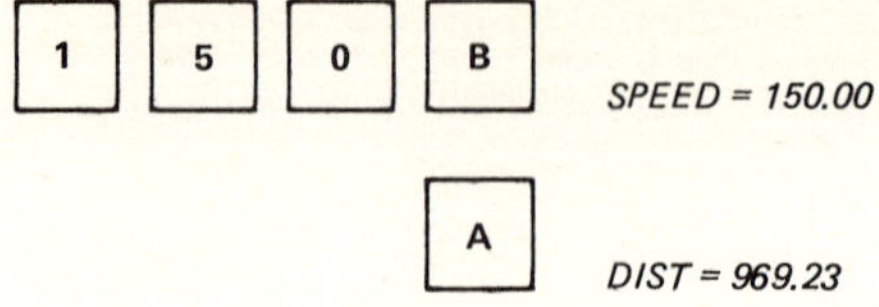

FIGURE 22-8 Range figured, using the
FLIGHT MANAGEMENT program in the
HP-41C.

With the TI59, having the endurance figure, we enter it into the
computer, using the appropriate keys to change it from the hour:
minute:second format to decimals, and then multiply it by the speed to
arrive at the no-reserve range (see Figure 22-7). It makes no difference
whether the speed is entered first, as in our illustration, and the endurance second, or vice versa.

The HP-41C still has all the applicable information with reference
to the endurance in its memory. All we now need to do is to enter 150
and identify it as the speed by pressing **B** (see Figure 22-8). Now
pressing **A** will produce the display showing *DIST=969.23*.

By the way, if you happened to turn the HP-41C off at any time
between performing these various time, speed, distance, and fuel calculations, you will find that the initial display is some series of digits
which coincide with the last value that was displayed before the computer was turned off, though, if that last figure was a time value, such
as 26:30:00, the display now appearing will be in the decimal format,
namely, *26.50*. You will then have to reinstruct the computer to use the
FLIGHT MANAGEMENT program by going through the keystrokes
shown at the beginning of the ETE problem: **XEQ APHA F M ALPHA**
which then restores the *D S T FF F* display.

23
DENSITY ALTITUDE

Density altitude is the altitude at which the airplane "thinks" it is, based on the actual (pressure) altitude and on the prevailing temperature. Because the performance of all airplanes and of nonturbocharged engines is significantly degraded by the reduction in air density resulting from increases in altitude and temperature, awareness of the density altitude is of considerable importance, especially when taking off from a high-altitude airport on a hot day.

For the determination of density altitude the standard calculator is completely useless, because I have never yet met a pilot who knows the mathematical formula which is used in arriving at the density-altitude figure.

The values which must be known in order to obtain the present density altitude are pressure altitude (the actual elevation of the airport) and temperature expressed in degrees Celsius. For the purpose of our examples I have chosen a pressure altitude of 2528 feet, and a temperature of 84°F.

The E6b can be used to determine density altitude, though the result is at best an approximation (see Figure 23-1). First we must use the scale at the bottom of the E6b to convert 84°F to degrees Celsius. As is seen in the illustration, that seems to be roughly 29°C. Now, using the inner portion of the E6b we place the pressure altitude on the center scale next to the temperature (**2600** and **29** in the illustration) and the indicator at the bottom will point toward the density altitude which appears to be more or less halfway between **4000** and **5000**. We therefore assume that the density altitude is about 4500 feet.

Density altitude calculations are where the electronic aviation computers really come into their own. Using the Avstar we first press

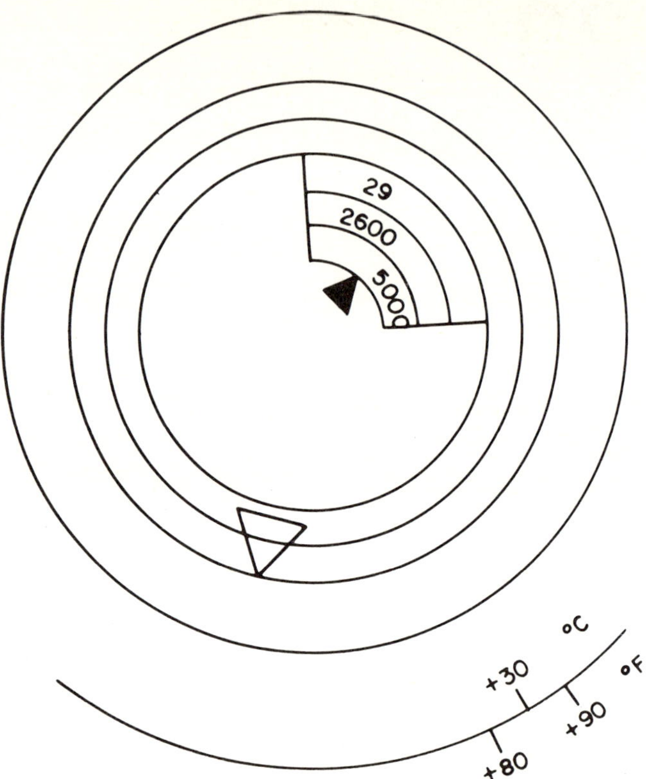

FIGURE 23-1   Density altitude can be figured with the E6b.

AS/Alt to identify the problem as one involving altitude (see Figure 23-2). We then enter the pressure altitude (2528) and identify it as such by pressing **P Alt**. We then enter the temperature in degrees Fahrenheit (84) followed by **Conv** and **°F°C** to convert that value to degrees Celsius. The display will show the Celsius value, *28.888889*. We identify that figure as the true temperature by pressing **T°C** and follow that with **Comp** and **D Alt** and the display now shows *4681* as the density altitude. It takes the computer a second or two to come up with that final result, an indication of the complication of the mathematic convolutions which are involved here.

The Commodore N60 has a density-altitude program which simply requires that we enter the pressure altitude and the temperature in degrees Celsius (see Figure 23-3). Since this computer insists on coming up with an endless series of decimal figures, the conversion of 84°F to degrees Celsius produces a display of *28.88888889*, and when the

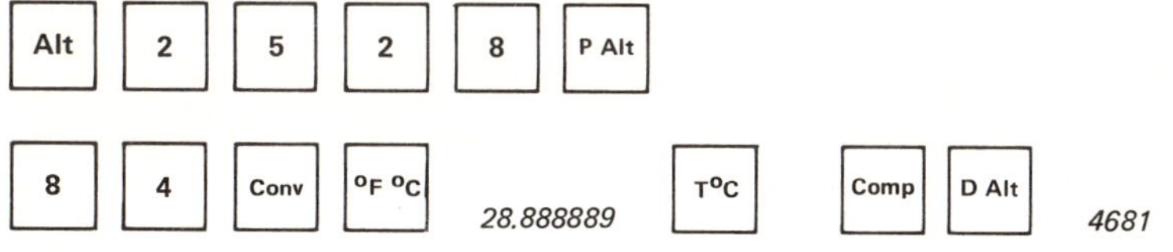

FIGURE 23-2   The Avstar has a special program to figure density altitude.

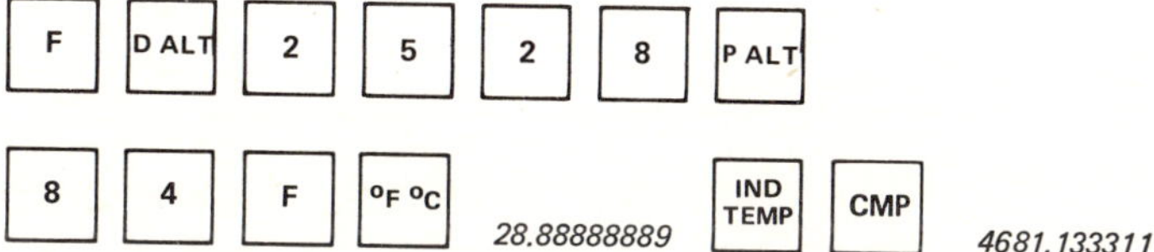

FIGURE 23-3   The N60, too, has a density-altitude program.

computation is completed the displayed result will be *4681.133311*, which is really somewhat more precision than we are after.

By the way, the instructions for figuring the density altitude which are printed in the instruction booklet show that CAS should also be entered. That is patent nonsense, since density altitude has nothing to do with the speed or even movement of the aircraft.

Using the Navtronic, the steps are slightly different (see Figure 23-4). First we press **D ALT** which results in the red light asking for **PRES ALT**. We enter 2528 and the red light now asks for **AIR TEMP °C**. We enter 84 followed by **CONV** and **°F-°C** and the display shows *28.888888*. We now press **ENT** and the display changes, after a brief hesitation, to *4697*. As can be seen, in this case the two computers don't agree completely. The discrepancy may be the result of the fact that while the Avstar rounded off the last temperature decimal to a 9, with the Navtronic it appeared as an 8. Such minor disagreements among computers are common, and are usually the result of the fact

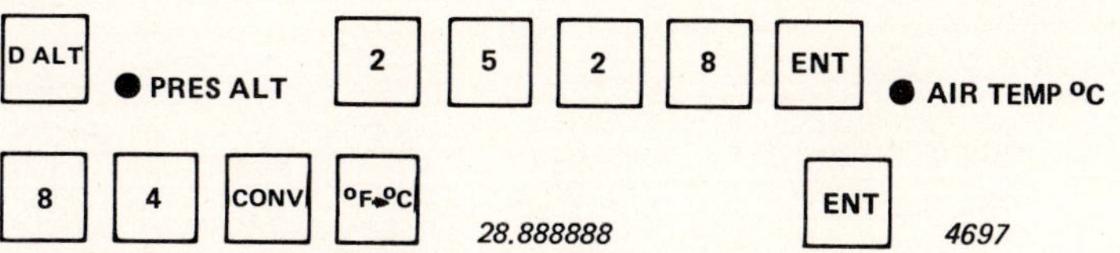

FIGURE 23-4   The Navtronic 1701 contains a program that figures density altitude.

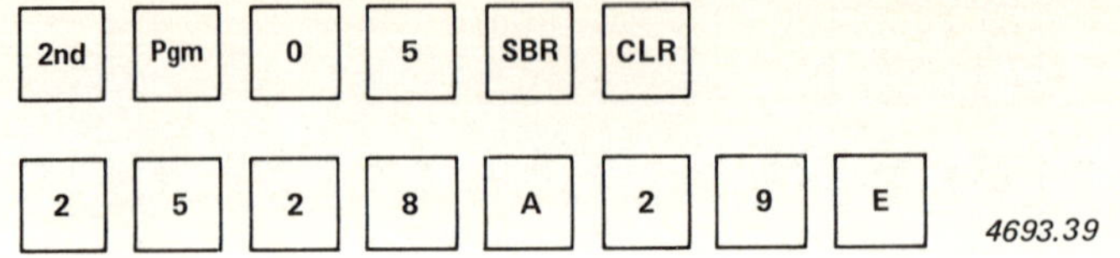

FIGURE 23-5   The TI59 uses the AV-05 program to figure density altitude.

that the display windows are limited in the number of digits which can be displayed. But the differences in the results are usually so small as to be of no practical consequence.

When using the TI59 (58C) we use the aviation program number 05: **2nd Pgm 0 5** and then initialize it by pressing **SBR CLR** followed by the airport elevation which we assign to key **A** and the temperature in degrees Celsius which is assigned to key **E**: **2 5 2 8 A 2 9 E** and the display comes up with the density altitude: *4693.39* (see Figure 23-5).

Neither the OC-1401 nor the HP-41C are included here because both seem to lack an easy-to-use density-altitude program.

24
TRUE AIRSPEED (TAS)

This is one of the problems which we face on every flight not just once, but every time we change altitude. While most of the newer airspeed indicators are equipped with a rotating scale which, supposedly, shows the TAS at the altitude to which it is set, the information obtained is quite inaccurate. Furthermore, in trying to determine the most economical altitude to choose for a flight under given wind conditions, we often need to know the TAS we can expect at different altitudes.

True airspeed is affected by three parameters: Altitude, temperature, and either indicated or calibrated airspeed (IAS or CAS). The mathematical calculations involved in deriving TAS from those three figures are quite complicated and not generally known by the average pilot. Therefore, the standard calculator is useless in trying to figure this problem. But it can be obtained by using the E6b, though, again, the result is a fairly rough approximation. To get precise results it is necessary to use any one of the aviation computers, all of which are programmed to provide this information.

In the examples below we have assumed that we are flying at an altitude of 10,500 feet, that the outside air temperature (OAT) is 67°F and that the CAS (or IAS) is 125 knots. (Since the CAS is rarely available, and since the difference between IAS and CAS is usually simply a matter of 2 or 3 knots, it is perfectly acceptable to use IAS, even though most computers call for an input of CAS.)

When using the E6b we first convert the degrees Fahrenheit to degrees Celsius by using the scale at the bottom. According to it 67°F is approximately the same as 20°C (see Figure 24-1). We now place **20** (temperature) over the altitude on the first and second scales on the

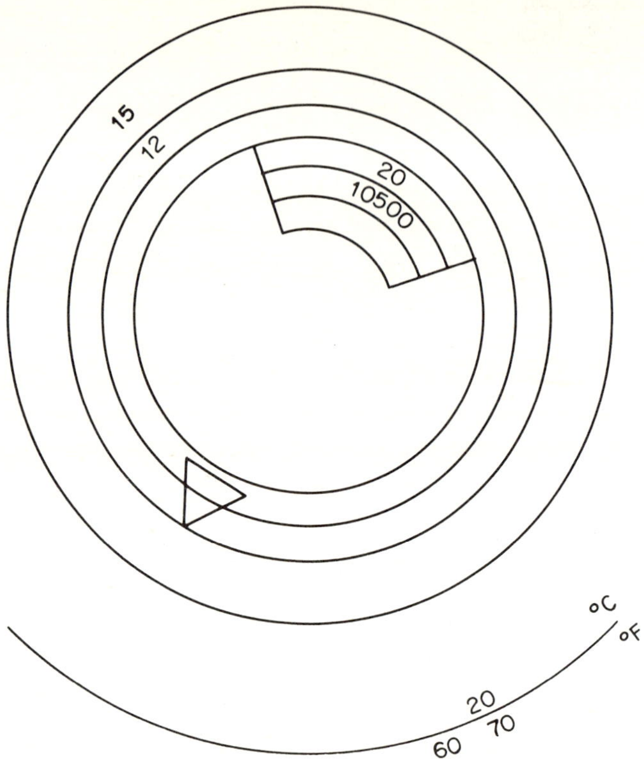

FIGURE 24-1   TAS figured with the E6b.

inside portion of the E6b. Then, by checking the relationship between the numbers on the second and first outer scales we find the IAS (or CAS) on the second scale (**12** which stands for 120) and the TAS on the first outer scale (**15** which stands for 150). Since we have assumed that our IAS is 125, it would seem that our TAS will be about 153 or 154 knots.

With the Avstar we first press **AS/Alt** to identify the problem of being one which involves airspeed and altitude (see Figure 24-2). We then enter the altitude (10,500) and identify it by pressing **P Alt**. We then enter the temperature 67 and convert it to degrees Celsius by using the **Conv** and **°F°C** keys and the display will come up with *19.444444*. We identify it to the computer, using the **T°C** key. Next we enter the TAS or CAS (125) followed by the **CAS** key. Then using the **Comp** and **TAS** keys the display will show *153* which represents the TAS.

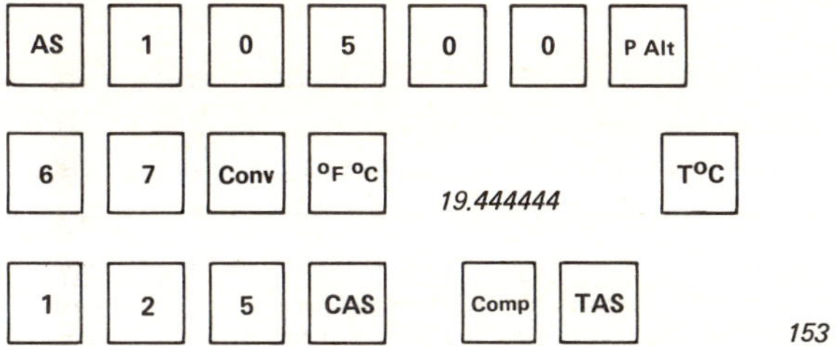

FIGURE 24-2   TAS figured using the Avstar.

As is always the case with the Avstar, the consecutive order in which this data is entered is of no consequence. We only have to be certain to identify each entered value by pressing the appropriate key after entry of the digits.

The Commodore N60 is equipped with a built-in TAS program which calls for the pressure altitude, the calibrated (or indicated) airspeed, and the temperature in degrees Celsius. With that information properly entered and identified, pressing **CMP** will result in a display of the TAS, again down to the nth decimal (see Figure 24-3). (The N60 has a ten-digit display and appears not to be happy unless all those digit positions are used.)

In calculating true airspeed on the OC-1401 we utilize one of its prestored aviation-program capabilities. We enter the CAS (125) by pressing **1 2 5 F CAS** (see Figure 24-4). (If CAS is not readily available, IAS will do.) We then enter the temperature (19°C) by pressing **1 9 °C**.

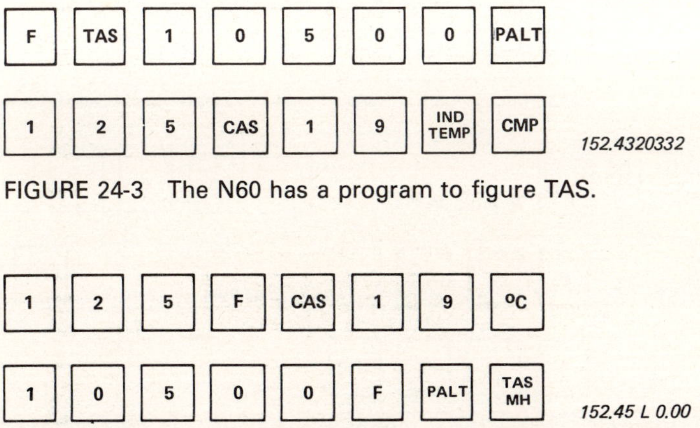

FIGURE 24-3   The N60 has a program to figure TAS.

FIGURE 24-4   The TAS program in the OC-1401.

Next we enter the pressure altitude (current altimeter reading with the altimeter set to 29.92) by pressing **1 0 5 0 0 F PALT**. Now by pressing **TAS MH** the display comes up with *152.45 L 0.00*. This means that the TAS is 152.45. The remaining portion of the display can be ignored. The program is set up to provide TAS as well as magnetic heading, but since no heading information was entered to start with, no such information could be computed. The small-size *L* which shows up in the display is actually not an L at all, but represents the sign ∠ which is used to designate that values are entered or displayed in terms of degrees.

With the Navtronic 1701tr we start out by pressing **TAS** which causes the red light to ask for **PRES ALT** and we enter it (10,500) resulting in a request for **AIR TEMP °C** (see Figure 24-5*a*). We now enter 67 followed by **CONV** and **°F-°C** and the display shows *19.444444*. With that value entered into the computer, it now asks for **CAL AIR SPD** and we punch it in (125) and the display, after a brief hesitation, will come up with *153*.

With the Navtronic it is also possible to figure out the CAS based on known TAS, altitude, and temperature. I can't imagine why anyone

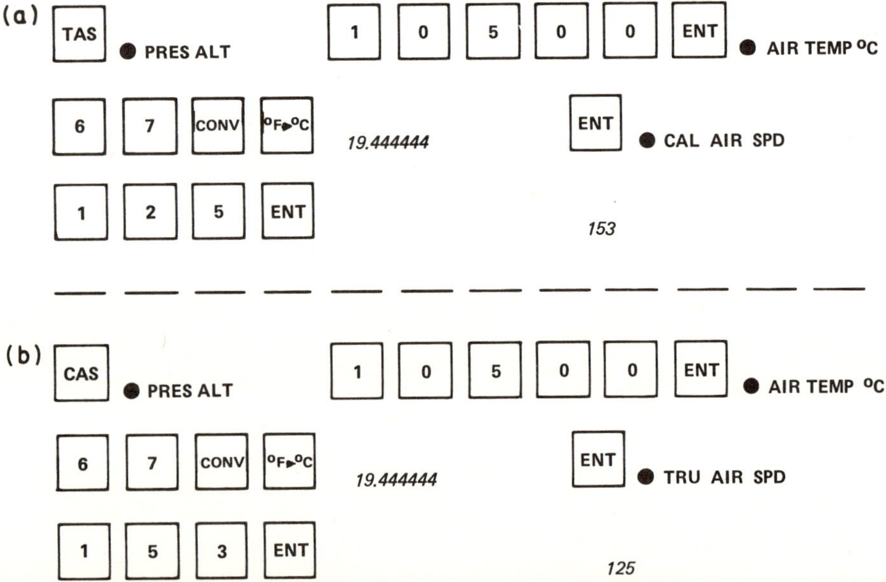

FIGURE 24-5   The program in the Navtronic 1701 can be used to figure either (a) TAS or (b) CAS.

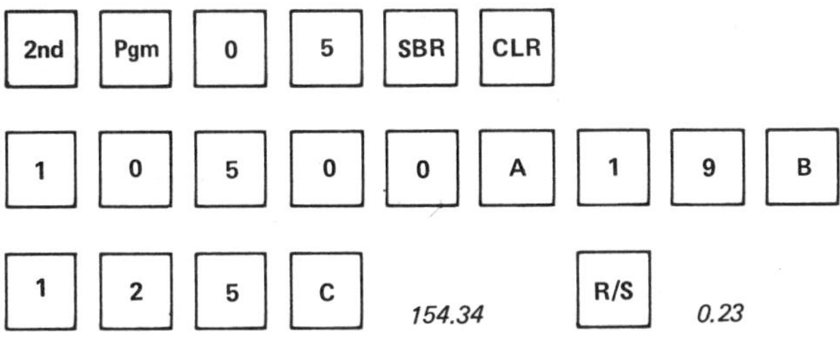

FIGURE 24-6   The TI59 uses the AV-05 program to figure TAS or Mach number.

would want to do this, but because it is available, I have shown it in Figure 24-5*b*.

To determine TAS the TI59 (58C) uses aviation program number 05. We select and initialize that program in the usual manner (**2nd Pgm 0 5 SBR CLR**) and then place the pressure altitude into the **A** position by pressing **1 0 5 0 0 A** and then place the indicated temperature in degrees Celsius into position **B** by pressing **1 9 B** followed by placing the indicated (or calibrated) airspeed into position **C** by pressing **1 2 5 C** which results in a display showing the TAS in knots: *154.34* (see Figure 24-6). If we now press **R/S** it results in the display changing to *0.23* which is the Mach number for that speed.

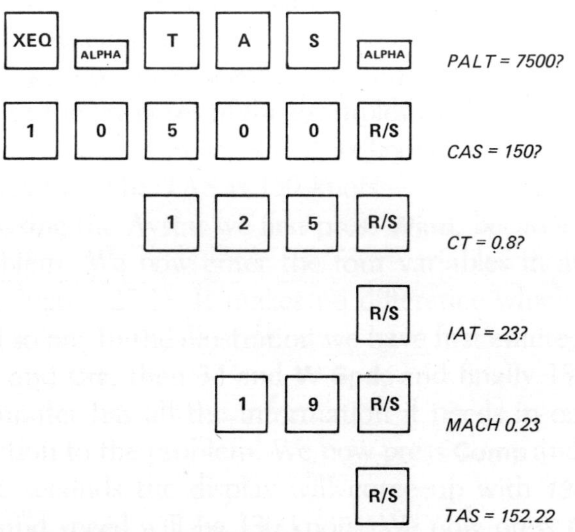

FIGURE 24-7   The HP-41C uses its TAS program to figure TAS or Mach number.

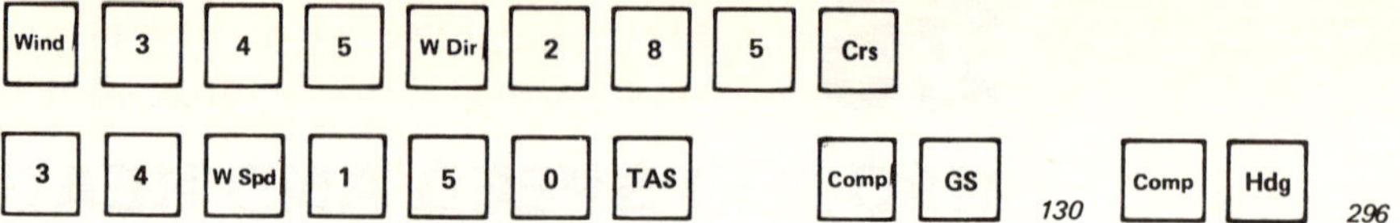

FIGURE 25-1 The Avstar can be used to figure ground speed and heading under given wind conditions.

with **296**, meaning that we have to fly a heading of 296° in order to stay on our 285° course.

It should be pointed out here that since the computer has no way of knowing where, on earth, it is located, it cannot make allowances for easterly or westerly variations. The trouble which arises is that winds are always reported in direction based on *true* north, while we, as pilots, do virtually all of our navigating with reference to the *magnetic* north. In order for any wind-related problems to provide correct answers, all directions must be in either *true* or *magnetic* terms. But, except in areas in which the amount of magnetic variation is extreme, the slight error caused by using *true* winds and *magnetic* courses is not sufficient to be of any serious consequence. After all, upper-wind forecasts are known to be approximations at best.

There is one more peculiarity with the Avstar which has to do with the solid-state circuitry employed. In cases where we are flying into a headwind which is less than 9° to the left of our intended course, the number 360 must be added to the wind direction. For instance, if our course is 285° and the wind is blowing from 280°, then, when entering the wind direction into the computer, we must enter 280 + 360 = 640. This is just one of those things that owners of Avstars must remember. That, at least, is what the instructions say. I ran the problem I described above, once adding that 360, and once not adding it, and each time the computer came up with the same answer (284° and 116 knots). So maybe it is not that important after all.

With the N60, just like with the other computers, we need to enter wind direction and wind velocity as well as the course and TAS, and, assuming each value has been identified correctly to the computer, the result, after pressing **CMP** to tell the computer to compute, will display the heading as **296-19-14**, meaning 296° 19′ 14″ (again that excessive amount of accuracy) and then, by pressing **GS** the display changes to the ground speed: **130.08161** (see Figure 25-2). Here, too, if

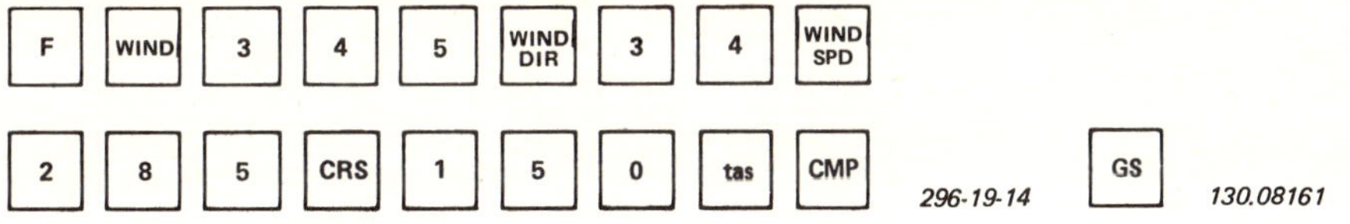

FIGURE 25-2   The N60 has a program designed to figure ground speed and heading under given wind conditions.

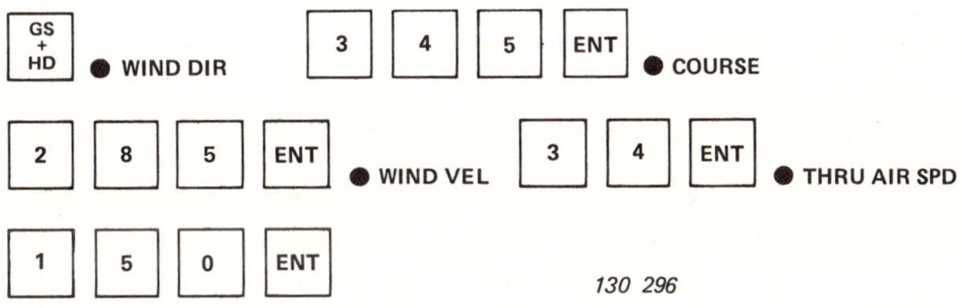

FIGURE 25-3   The Navtronic 1701 makes determining ground speed and heading under given wind conditions easy.

wind direction is given in true north, and the course in magnetic north, the result will be slightly off, because the computer is not programmed to allow for magnetic variations.

To work the problem with the Navtronic 1701tr we start by pressing **GS+HD** which produces the request for **WIND DIR** (see Figure 25-3). We enter 345 and the computer now asks for **COURSE**. We enter 285 and are asked for **WIND VEL**. We enter 34 and are requested to provide **TRU AIR SPD**. We enter it (150) and the display window now shows *130 296*, the first value being the ground speed, and the second value the heading.

With reference to true and magnetic directions, the same holds true here that was talked about above with reference to the Avstar and N60.

Neither the OC-1401 nor the TI59 nor the HP-41C contain an easy-to-use program for this problem. In each instance it is a part of a much longer and more complicated program designed to provide all answers relative to an entire flight. These programs are explained in some detail in Chapters 7, 9, and 10.

26

DETERMINING WIND DIRECTION AND VELOCITY IN FLIGHT

Every once in a while, especially when we hold a considerable crab angle and find that it seems to take inordinately long to get from one navigation aid to the next, we'd kind of like to know what kind of wind components we are dealing with. Here the problem is exactly the opposite of the one we have described in Chapter 25. In this case we know the heading we are holding in order to stay on course, and we know the ground speed, either from an on-board DME or from the time it took to get from one VOR to the next. If we give that information to the computer, it will come up with the answer.

For the examples I have used the figures which were the result of the previous computation: Heading is 296°, ground speed is 130 knots, and course and TAS remain the same (285 and 150).

With the Avstar we press **Wind**, then enter the heading (296) and press **Hdg**. We then enter the course (285) and press **Crs**. We now enter the ground speed (130) and press **GS** followed by the true airspeed (150) and **TAS**. We then use the **Comp** and **W Spd** keys and after some hesitation the display will come up with *33* representing the wind velocity (see Figure 26-1). Obviously somewhere in the mathematical calculations performed by the computer 1 knot has gotten lost. But that really doesn't matter much. We now press **Comp** and **W Dir** and the computer has decided to display *344* which is again 1° short of the previous figure. Well, I'm sure we can live with that.

In order to determine the direction and velocity of in-flight winds, the N60 needs to know the TAS, the ground speed, the course, and the heading. With this information properly identified to the computer, pressing **CMP** will produce the wind direction in degrees, minutes, and seconds, and then pressing **WIND SPD** the wind speed will

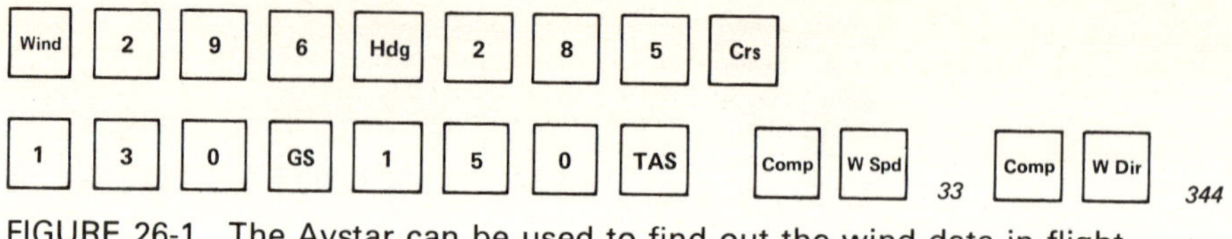

FIGURE 26-1   The Avstar can be used to find out the wind data in flight.

appear (see Figure 26-2). Assuming that course and heading were given in magnetic degrees, the wind direction will also be in degrees magnetic.

Using the Navtronic 1701tr we start by pressing **WV+WD** and the red light calls for **HEADING** (see Figure 26-3). We enter it (296) and are asked for **COURSE**. We now enter that (285) and the computer wants to know **SPD**, which, in this case, means the ground speed (130). Having added that we are now asked for **TRU AIR SPD**. We enter this last value (150) and after a brief hesitation the display will come up with *33 344* representing the velocity and direction of the wind. Apparently the two computers use more or less the same circuitry, as both came up with the same answer.

In order to determine the wind direction and velocity in flight, the TI59 (58C) includes an aviation program, labeled 08, which can be used for this purpose with relative ease. It requires all the same information needed by the other computers, plus the magnetic variation

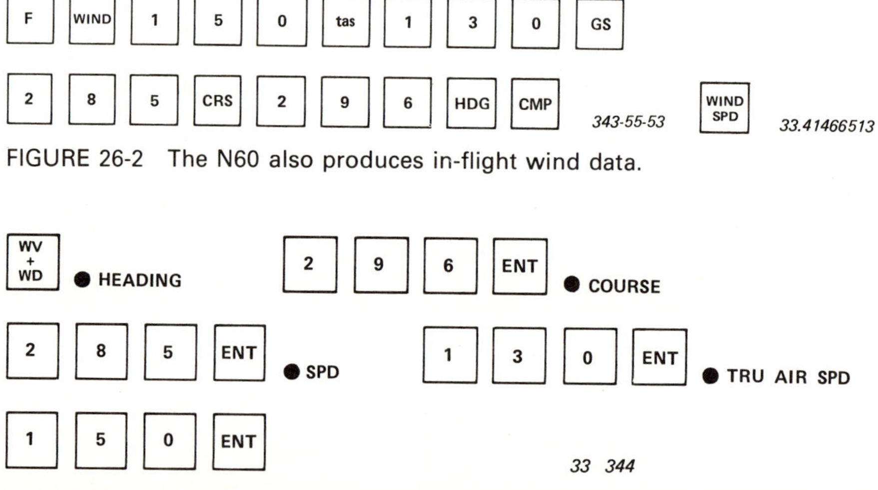

FIGURE 26-2   The N60 also produces in-flight wind data.

FIGURE 26-3   The Navtronic 1701 has a program to determine wind direction and velocity in flight.

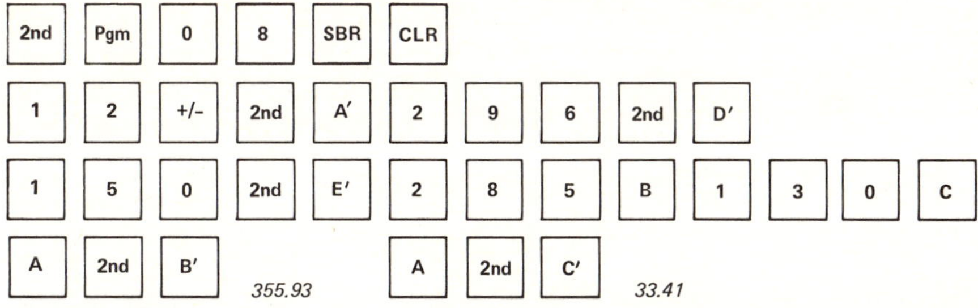

FIGURE 26-4   The TI59 uses a portion of its AV-08 program to determine in-flight wind data.

which is applicable to the aircraft's position, because it takes into account that course and heading are given in magnetic degrees, while wind directions are always reported in true degrees.

We first select the program by pressing **2nd Pgm 0 8** and then initialize it by pressing **SBR CLR** (see Figure 26-4). Next we enter the appropriate data and place it into the **A** through **E** and **A'** through **E'** positions: **1 2 +/− 2nd A'** (this puts the easterly variation into position **A'**) then **2 9 6 2nd D'** (putting the magnetic heading into position **D'**) then **1 5 0 2nd E'** (placing the TAS into position **E'**) then **2 8 5 B** (which puts the magnetic course into position **B**) then **1 3 0 C** (placing the ground speed into position **C**). Now, to calculate the wind direction

```
            XROM "IFW"
VAR=12?
                    RUN
TAS=150?
                    RUN
MC=040?
                    RUN
T AT WP1=?
          12.0000   RUN
D TO WP2=?
              47    RUN
T AT WP2=?
          12.1848   RUN
STEERING=?
             023    RUN
WIND=314.044
```

FIGURE   26-5 The thermal printout obtained by running the IN-FLIGHT WIND program with the HP-41C.

we press **C 2nd B'** and the display reads *355.93*, the direction of the wind in true degrees. We then press **A 2nd C'** and the display comes up with *33.41*, the velocity of the wind in knots.

For the HP-41C I have deviated from the format in the illustrations used so far, and have instead used the printout produced by the thermal printer which is part of the HP-41C system. The computer contains an IN-FLIGHT-WINDS program identified as IFW. To activate this program we press **XEQ ALPHA I F W ALPHA** and the display will come up with a request for information, while the printer, if attached, will print out the name of the program (**XROM "IFW"**) followed by the request which is for the applicable magnetic variation in the area in which we happen to be flying. Since the last variation we used with the computer was 12° westerly, the display (and the printout) shows *VAR=12?* Assuming that is correct, we can go to the next step. If another variation must be entered, we simply press the appropriate numerical keys, and the display will show the new number without the identifying *VAR=*. Now to go to the next step we press **R/S** which stands for *run/stop*. The computer now comes up with a request for the TAS, showing the last TAS which had been used previously. In our example that last figure was 150 and the display as well as the printout therefore show *TAS=150?* Between the two lines the printout also shows the word **RUN** to show that the **R/S** key has been used. Assuming that we are satisfied with the TAS figure (we can change it in the same manner described above), we again press **R/S** and the next request deals with the magnetic course. Since the last course entered was 040, the display shows *MC=040?* Assuming this is fine, we press **R/S** to go to the next step. Now, instead of requesting the ground speed, the computer wants to know the time at which we crossed a checkpoint, referred to here as *waypoint 1*. The display shows *T AT WP 1=?* Let's say that we crossed that waypoint at 12 noon. So we enter 12.0 and while the display changes to the next question, the printer prints out **12.0000**. The computer now displays *D TO WP 2=?*, meaning that it wants to be told the distance between waypoint 1 and waypoint 2. Here we have assumed a distance of 47 nautical miles, and we have entered that by pressing the appropriate numerical keys. The next question shows up as *T AT WP 2=?*, meaning that the computer wants to be told the time at which we crossed waypoint 2. Since it took us 18 minutes and 48 seconds to fly that distance, we now punch in 12.1848 which then appears on the printout as well as in the display. Pressing **R/S** now produces *STEERING=?*, which means that the computer now needs to know the magnetic direction in which we are steering, in other words, the heading. Here we entered our heading as 023° which appears in the display as well as the printout. Now pressing **R/S** again results in a few seconds' hesitation before the display comes up with *WIND=313.044* which identifies the wind direction as 313° *true* and the wind velocity as 44 knots.

The computer program is designed to deal with combinations of true and magnetic degrees, which is why the first question asked is about the magnetic variation, which must be included in the information in order to permit the computer to accurately perform its calculations.

27
RATE OF CLIMB BASED ON GRADIENT AND DISTANCE

There are times when it is important to be able to figure out the rate of climb (ROC) in feet per minute, using other types of information. One such instance involves instrument flight rules (IFR) approaches or departures. On many IFR charts we are likely to find information which calls for an altitude change of so many feet per mile, a value which is difficult to translate into feet per minute. (It should be pointed out here that it makes no difference whether we talk about the rate of climb or the rate of descent. The mathematics are the same.)

The mathematical formula employed in transposing feet per mile into feet per minute calls for dividing the number of feet by 60 and then multiplying the result by the ground speed. This can be done with any standard calculator or with the calculator portion of any of the portable aviation computers.

The other ROC-related problem involves finding the required rate of climb in order to achieve a given altitude change within a certain distance. This would occur, for instance, if we took off toward the west from Denver and needed to know what rate of climb would be necessary in order to clear the Rockies which would be staring us in the face.

The examples illustrate how to solve both those problems with the standard calculator and the different aviation computers. In the upper portion of each illustration we have assumed that we need to achieve an altitude change of 285 feet over a 1-mile distance while flying at a ground speed of 115 knots. (In this problem the speed must be given in knots and not in miles per hour in order for the result to be accurate.) In the lower portion we have assumed that we are cruising at 8730 feet and that we want to climb to 14,000 feet in 11 miles at a ground speed during climb of 130 knots. (In this case distance and speed can be either in nautical miles and knots or statute miles and miles per hour.)

Using an ordinary calculator, here is what needs to be done: To solve the first problem we simply divide 285 (the altitude change in feet per nautical mile) by 60 which will produce a display showing *4.75* (see Figure 27-1). We now multiply that by the ground speed (115) and the display shows *546.25* which represents the rate of climb or descent in feet per minute to achieve an altitude change of 285 feet in 1 nautical mile.

The second problem involves two steps. First we enter the altitude to which we want to climb (14,000) and deduct from it our current cruising altitude (8730) to arrive at the amount of altitude change required, namely *5270* feet. We must now make a note of that or remember it. We now start again by entering the distance in which the altitude change must be accomplished (11) and divide that by our ground speed (130) and the display will show *.08461538* which represents the fraction of an hour which it will take to cover those 11 miles. But since ROC is expressed in feet per *minute* (not *hour*) we now multiply that figure by 60 and the display will change to *5.0769228* which identifies the number of minutes it will take to cover that distance. We now store that value in the memory and then enter the

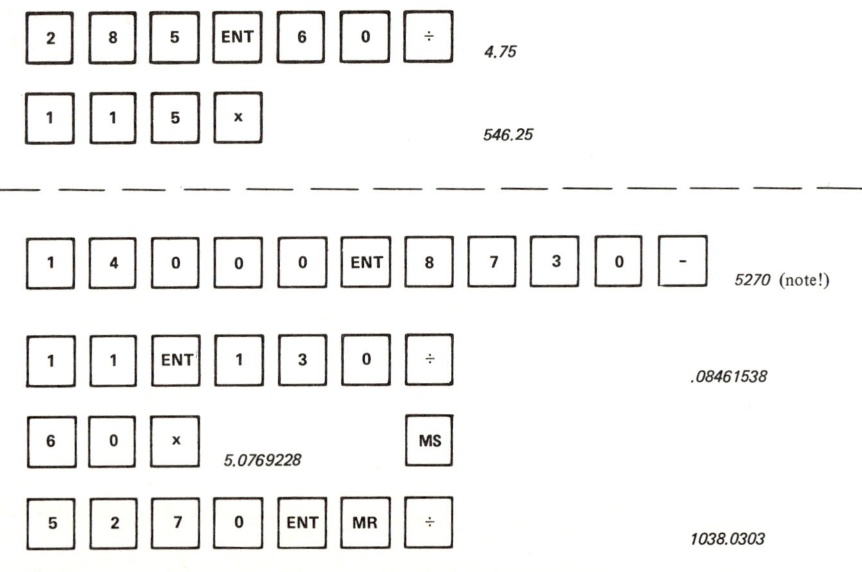

FIGURE 27-1   Rate of climb figured with the calculator.

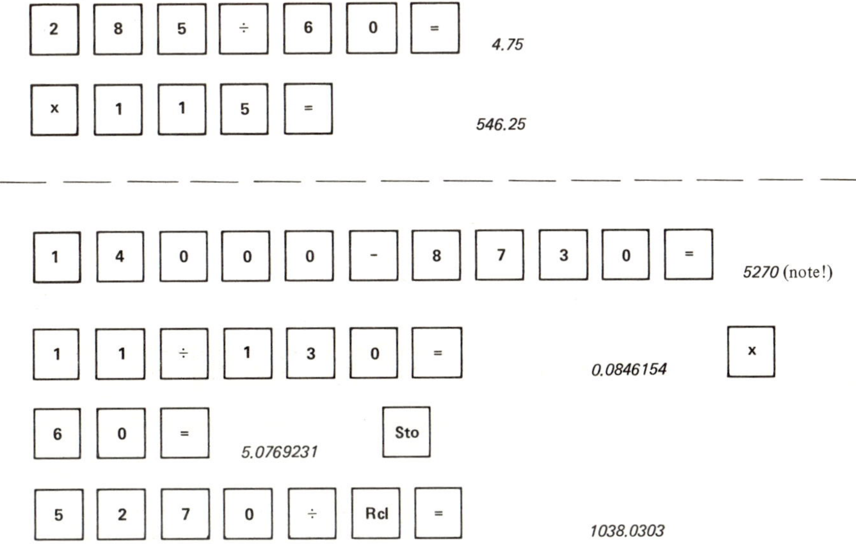

FIGURE 27-2   The Avstar is used here to figure rate of climb.

amount of altitude change (5270) which we have either remembered or written down. Now we recall the previous figure from the memory and ask the calculator to divide one by the other. The resultant display shows *1038.0303* which represents the number of feet per minute which we must be able to climb in order to achieve that altitude change within that distance.

With the Avstar both problems are solved using simply the calculator portion of the computer, and the steps involved are exactly the same as those described for the calculator (see Figure 27-2).

Like the Avstar, the N60 doesn't have a program to solve this problem. It must be accomplished by using the calculator portion and the memory-storage capability.

To determine the rate of climb in feet per minute based on the number of feet of altitude change in a nautical mile, we divide that number of feet (285) by 60 and multiply the result by the ground speed (115) and the required rate of climb in feet per minute is displayed as *546.25* (see Figure 27-3).

To determine the rate of climb required to achieve a given altitude change within a given distance, we first determine the altitude change by deducting the current level of flight from the level of flight to which we want to climb (14,000 minus 8730) and we then store the result

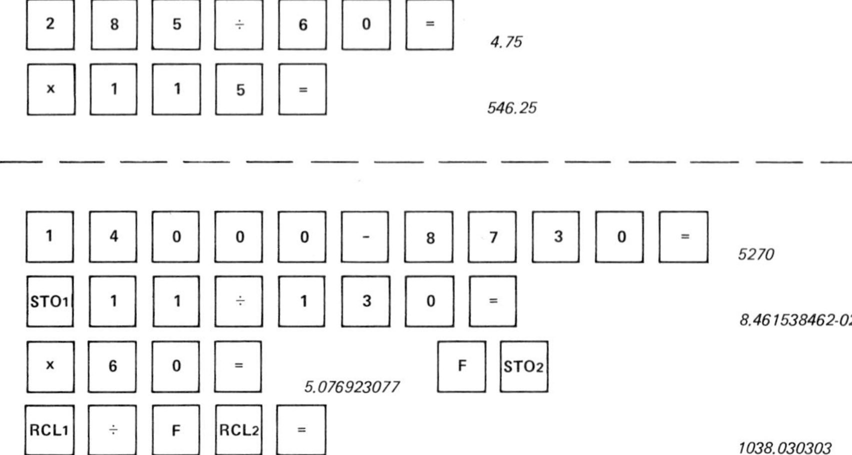

FIGURE 27-3   Using the N60 in figuring rate of climb.

(5270) in memory number one by pressing **STO₁**. We then divide the distance available for the climb (11) by the ground speed (130) and multiply the result (8.461538462-02) by 60 and store that new result (5.076923077) in the number two memory by pressing **F STO₂**. We now recall the value stored on the number one memory by pressing **RCL₁** and divide it by the amount stored in memory number two, producing a display of *1038.030303* as the required rate of climb to get from 8730 feet to 14,000 feet in 11 miles at a ground speed of 130 knots.

About that crazy *8.461538462-02* readout, the *–02* means that the decimal point should actually be two spaces to the left of where it is shown, which is why multiplying that figure by 60 results in a figure which appears to be smaller than the original figure.

With the OC-1401, to determine the rate of climb based on gradient information, meaning the number of feet of altitude change which must be accomplished in a mile (or any other given distance) we enter the altitude change in feet (285), divide it by 60, and multiply the result by the ground speed (115) to arrive at the rate of climb in feet per minute: *546.25*. (See Figure 27-4.)

To determine the required rate of climb based on the distance available to achieve a given altitude change, we first deduct the current flight level (8730) from the new altitude (14,000) and then store the result (*5270.00*) in memory number one. We then enter the distance available to achieve that altitude change (11) and divide it by the ground speed expected during climb (130), and we divide that result by 60 to arrive at the time in decimal format which will be spent in covering the available distance. We store that result (*5.08*) in memory number two. We now divide the value stored in memory number one by the value stored in memory number two and the display shows that we will have to climb at *1038.03* ft/min in order to get up there within the given distance.

With the Navtronic 1701tr both of these problems can be solved somewhat more simply (see Figure 27-5). For the first one we press **ROC** and are asked to enter **DIST**. We punch in 1 because we're talking about 1 nautical mile. Now the computer wants to know **SPD**. We enter 115 and are asked to provide **ALT CHG**. We enter 285 and the display will show *546.3* as the required rate of climb or descent.

For the second problem we again start with **ROC** followed by **DIST** (11) followed by **SPD** (130) followed by the request for altitude change (**ALT CHG**). We now punch in the altitude to which we want to climb (14,000) and deduct from it our current cruise level (8730) which results in a display showing *5270*. We now press **ENT** to enter that into the computer and the display changes to *1038.0*, the rate of climb in feet per minute needed to get up to the new altitude in 11 miles.

To determine the rate of climb needed to achieve a given climb gradient, as shown on an approach plate, in this instance, to climb 285 feet in a mile, the TI59 (58C) uses aviation program 18. We enter and initialize it in the usual manner (see Figure 27-6). We then enter

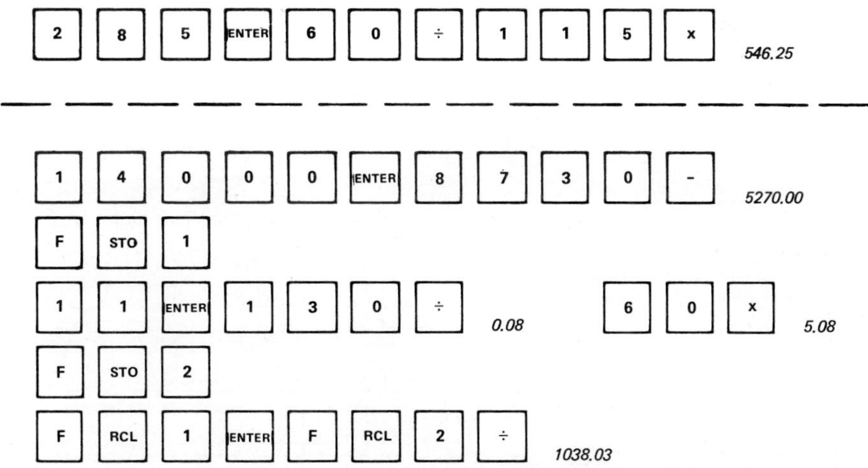

FIGURE 27-4   The OC-1401 uses two of its memories in figuring rate of climb.

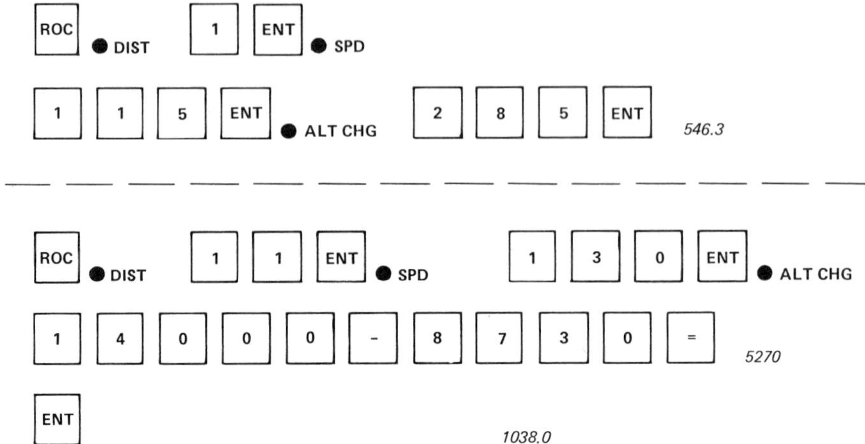

FIGURE 27-5   The Navtronic 1701 has a program designed to figure rate of climb.

the altitude change (285) and place it into position **A'** by pressing **2nd A'**. We then place the ground speed (115) into position **B'** in the same manner and then enter the distance in miles (1) and place it into position **A**. Now, when we press **B** the display will show *545.6501*, representing the necessary rate of climb to achieve a 285-foot altitude change over a 1-mile distance at a ground speed of 115 knots.

To determine the rate of climb needed to get over a mountain range or some such, we use actually the same routine, using the same

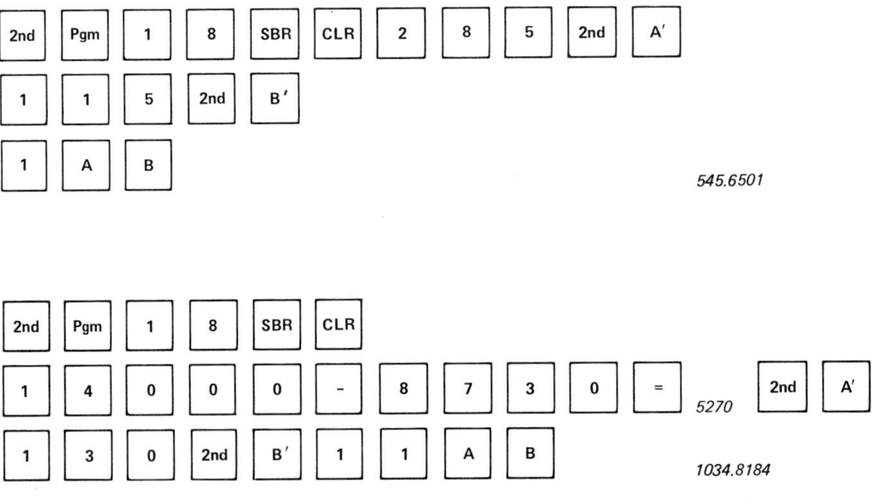

FIGURE 27-6   The TI59 uses the AV-18 program to figure rate of climb.

aviation program. We identify and then initialize the program. We then enter the desired altitude and deduct from it our current flight level, entering that result into position **A′**. We then enter the ground speed which is expected during the climb into position **B′** and follow that by entering the horizontal distance available for the altitude change into position **A**. Now, when we press **B** the display comes up with *1034.8184* as the required rate of climb.

28

AREA NAVIGATION

Some aviation computers can be used to perform certain simplified area-navigation (RNAV) functions. It goes without saying that in no case can this be called true RNAV, since real RNAV requires radio reception from one or several VOR/DMEs or VORTACs (VOR tactical air navigation). But, by using a number of different methods, these computers can provide us with the direct off-airways course between any two points even if they are considerable distances apart, plus the actual distance from point to point. By then using this information in conjunction with the program which determines heading and ground speed under known wind conditions, we are put into the position to fly a fairly precise direct off-airways route.

The N60 can be used in two different ways to determine RNAV direct routes. Using the same basic information which is used in conjunction with the Navtronic, the N60 works slightly differently.

First we identify the program, which is referred to as *navigation by reference to one VOR*, by pressing **F N1VOR** (see Figure 28-1*a*) after which we enter the distance to the waypoint (shown as Y on Figure 28-1*b*) by pressing **1 7 8 DIST₁** followed by the radial to that same waypoint: **6 8 R₁**. We then enter the distance to the other waypoint (X in Figure 28-1*b*) as **6 3 DIST₂** followed by the radial **3 3 6 R₂** and then press **CMP** which, after a fairly long "thinking" interval produces *87-15-34*, meaning that the course from X to Y is 87°15′34″. We then press **DIST** and the display changes to *190.8814447*, representing the distance in nautical miles between X and Y.

The other RNAV program is referred to as *rhumb-line navigation*. It uses latitudes and longitudes instead of VORs and is therefore more useful for long-distance flights and flights in parts of the globe where VORs are not available with any degree of frequency.

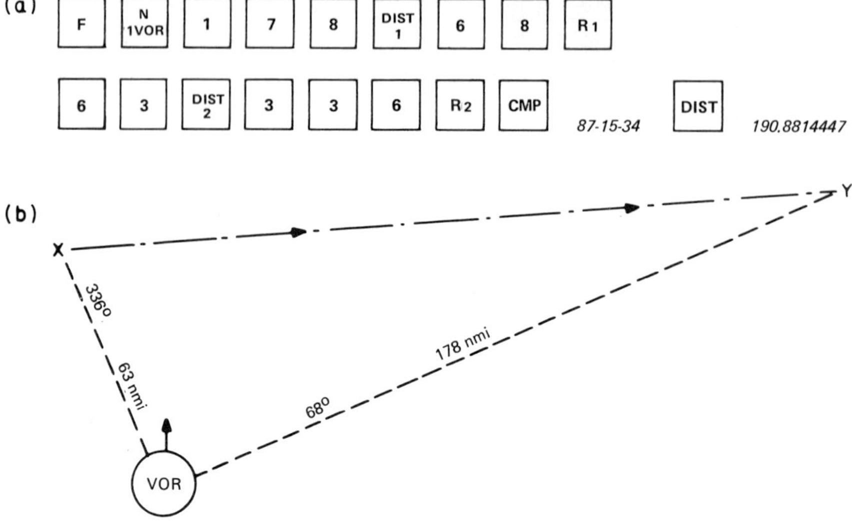

FIGURE 28-1 The N60 has an RNAV program (a), using data obtained from one VOR (b).

To activate the rhumb-line program we use these keystrokes: **F NRMB** and then enter the latitude in degrees and minutes (see Figure 28-2). In the illustration we have picked a flight from Chicago to Los Angeles. We entered the latitude for the first waypoint after leaving the Chicago standard instrument departure (SID), which is 41°16′. This is accomplished by pressing **4 1 HMS 1 6 F LATs**, this latter referring to the *starting latitude*. We then press **8 7 HMS 4 8 F LNGs**, to enter 87°48′ as the *starting longitude*. We repeat the procedure for the *destination latitude* and *longitude* in the same manner: **3 3 HMS 4 7 F LATd 1 1 8 HMS 0 3 F LNGd** and then press **CMP** which produces, again after some hesitation, the display *253-39-03*, meaning that the rhumb-

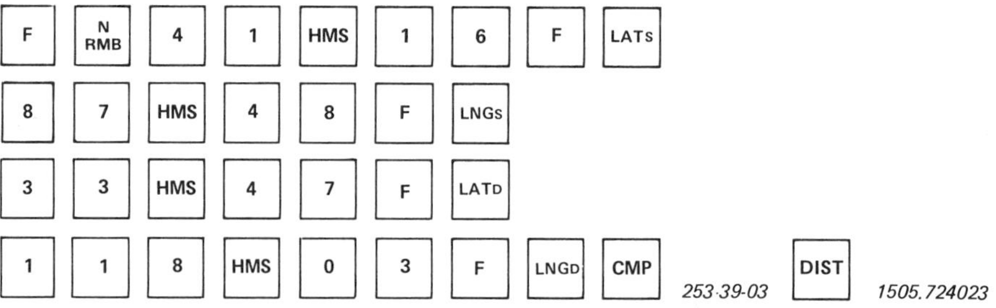

FIGURE 28-2 The N60 also has a rhumb-line program for flights involving greater distances.

line direction (not the great-circle route) from Chicago to Los Angeles calls for 253°39′3″. Then, if we press **DIST** the display changes to *1505.724023*, representing the straight-line distance between the two cities in nautical miles.

The OC-1401 uses two VORs in order to establish the magnetic course and distance for an RNAV direct route between two off-airways points. In the example used in Figure 28-3, the pilot wants to fly from point X to point Y, and wants to know the magnetic course to make good and the distance which has to be covered.

To accomplish this the pilot first feeds the distance and the radial from VOR1 to VOR2 into the computer by pressing **2 0 0 ∠ 3 0 0 F VR$_{12}$** (see Figure 28-3*a*). Then enter the radial and distance from VOR1 to the destination (point Y in Figure 28-3*b*) by pressing **8 0 ∠ 8 5 VR$_{1\triangle}$**. Then enter the radial from VOR1 to the departure point (point X in Figure 28-3*b*) by pressing **2 8 0 F RAD 1**. Last, enter the radial from VOR2 to the departure point by pressing **2 1 0 RAD 2**. Now the computer has all the information it needs in order to perform the necessary computations. We now press **DST MH** and for a few moments the display will show *L1* (leg 1) to let us know that the computer is "thinking." And then the display changes to *290.85 L 95.91*, meaning that the distance is 290.85 nautical miles and the magnetic heading from X to Y is 95.91°. If, for some reason, we now want to change that latter value into degrees, minutes, and seconds we first have to clear the display by pressing **CLR X** and then enter the degrees again (95.91). We then press **F H** and the display shows that the magnetic heading is *95.54.36* or 95°54′36″.

The Navtronic 1701r and 1701tr contain an RNAV capability which uses the radials and distances of two points from one VOR (as is shown in Figure 28-4*b*). To determine the course which must be flown to get from point X to point Y we first press **CONV** and **RNAV** (see Figure 28-4*a*). The red light now asks for **R2**, meaning the radial from the VOR to point Y. We enter that value (68) and the computer now asks for **R1**. We enter the radial to point X (336) and the computer now wants to know **D1**, or the distance to point X. We enter it (63) and are asked for **D2**, the distance to point Y. We enter that (178) and the display, after several seconds' delay, will come up with *191 087*, of which 191 represents the distance from X to Y, and 087 represents the *magnetic* course. (The results are always in terms of magnetic directions, since all VORs are aligned to the magnetic north.)

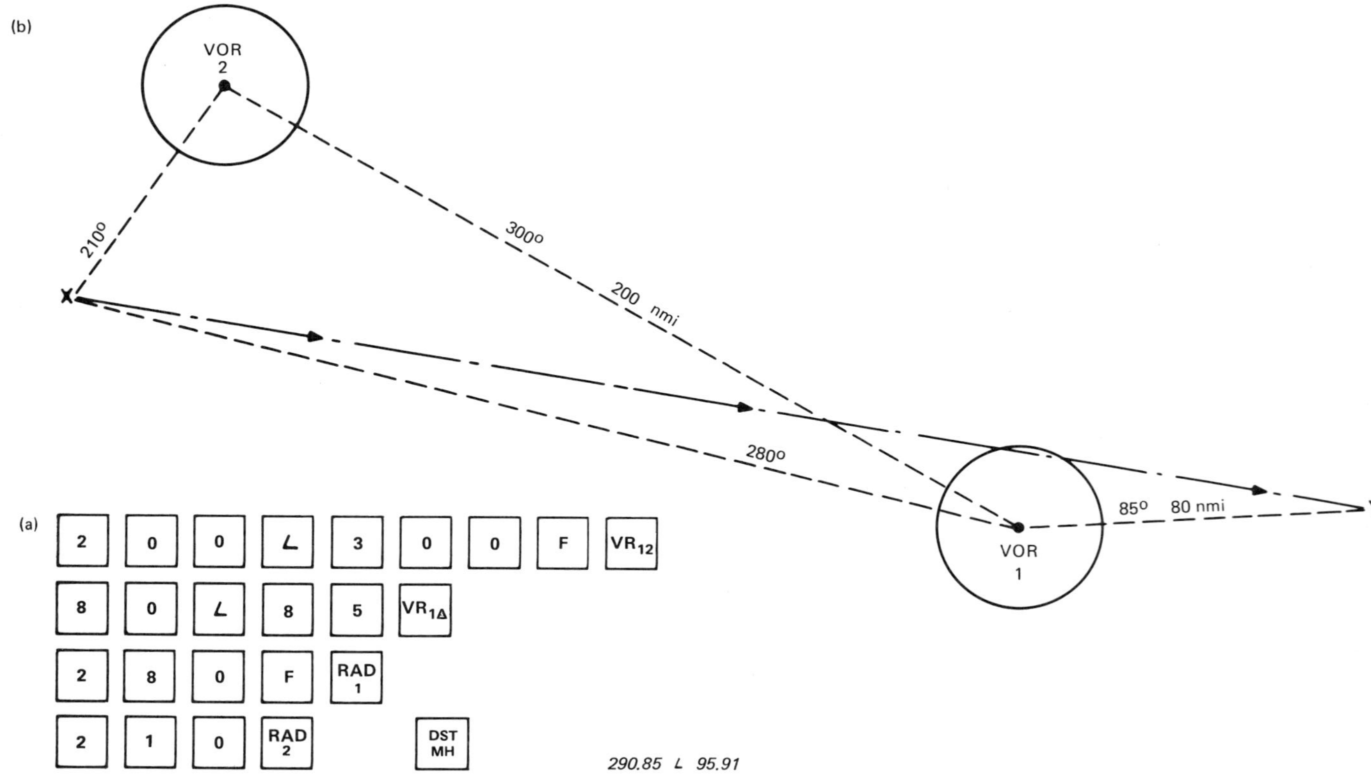

FIGURE 28-3   The OC-1401 has an RNAV program (a) using data relating to two VORs (b).

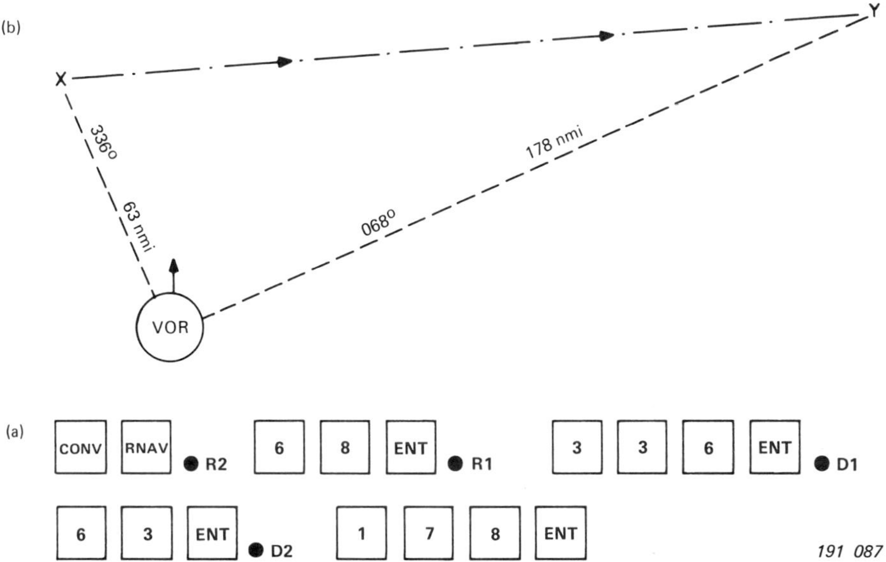

FIGURE 28-4   The Navtronic 1701r and 1701tr contain an easy-to-use RNAV program (a), using data relative to one VOR (b).

On the subject of RNAV the TI59 (58C) offers two programs, one based on the principle of rhumb-line navigation, meaning a straight line between two points (in accordance with a Mercator projection), and the other the great-circle route, the actual shortest distance between two points, a means of navigation which is useful only for very long flights.

For rhumb-line navigation the applicable program is aviation program 10. Running this program requires that we know the latitude and longitude figures for our departure and destination points. In Figure 28-5*a* we have again used a flight from Chicago to Los Angeles for our example. The answers we are looking for are the magnetic and true course and the total distance along that straight line.

We identify and initialize the program in the usual manner (see Figure 28-5*a*). We then enter the average magnetic variation, in our example −9° which happens to be the variation for Wichita, which is more or less halfway between the two points, by pressing **9 +/− 2nd A′**. Having placed the variation into position **A′**, we now enter the latitude for Chicago and place it into position **B** by pressing: **4 1 . 1 6 B** followed by putting the longitude into position **B′** by pressing **8 7 . 4 8 2nd B′**. Next we place the latitude and longitude for Los Angeles into posi-

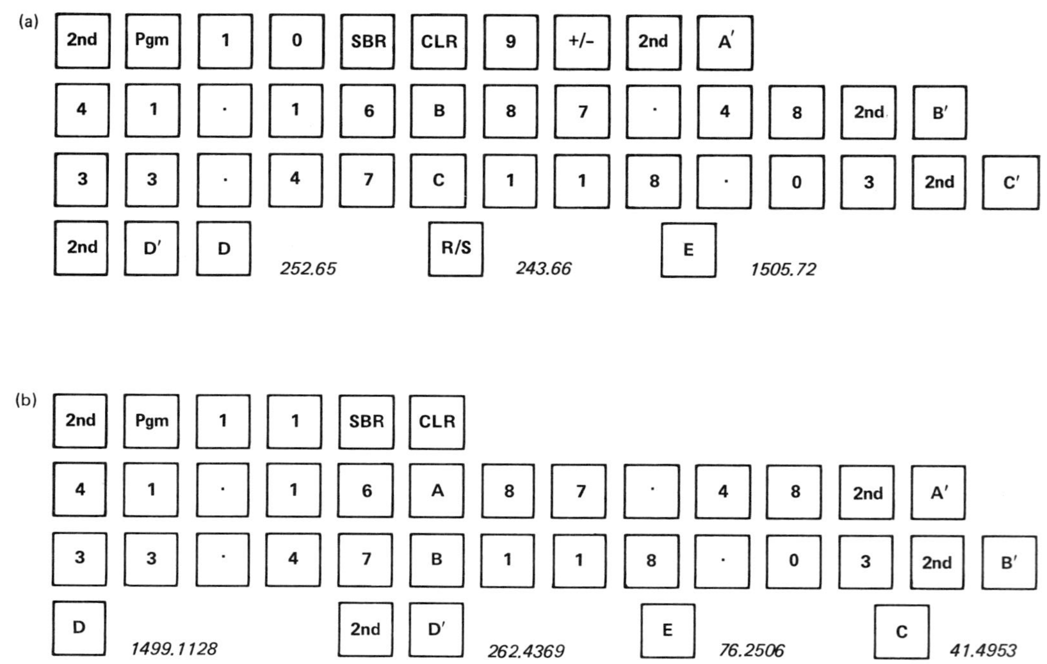

FIGURE 28-5   The TI-59 has two RNAV programs, both designed primarily for flights of greater distances. The AV-10 program (a) determines the rhumb-line route and distance between two points. The AV-11 program (b) computes the great-circle route and distance between two points.

tions **C** and **C′** by pressing **3 3 . 4 7 C 1 1 8 . 0 3 2nd C′**. We then press **2nd D′ D** and the display comes up with the true course: *252.65* degrees. Now, if we press **R/S** the display changes to *243.66* which is the magnetic course. We now press **E** and the total distance in nautical miles is displayed as *1505.72*.

Now let's use the great-circle navigation program for the same trip from Chicago to Los Angeles (actually too short a distance to make great-circle navigation meaningful, as we'll shortly see). The program used for this purpose is aviation program 11. We identify and initialize it in the usual manner (see Figure 28-5*b*). We then forget about the magnetic variation and start right out by entering the latitude and longitude parameters for the two cities into the appropriate key positions: **4 1 . 1 6 A 8 7 . 4 8 2nd A′ 3 3 . 4 7 B 1 1 8 . 0 3 2nd B′**. Now we press **D** and the display comes up with the great-circle distance: *1499.1128* nautical miles. Then, by pressing **2nd D′** the display changes to *262.4369* which is the initial true course. But since any great-circle route is, in fact, a curved line (a portion of a circle), flying the actual route would require constant heading changes. The practical alternative is to pick one of several points along this line (vertexes) and designate them as waypoints, and then fly a straight line between each of these waypoints. In our illustration, pressing **E** will produce the longitude of a point approximately halfway along the route, and pressing **C** then produces the latitude of that same point. Now, by using the rhumb-line program again, we can determine the straight-line courses and distances between the departure point and the waypoint, and then between the waypoint and the destination.

When running either of these programs you will notice that even though you have entered the latitude and longitude figures in terms of degrees and minutes, once they are placed into their appropriate key position, the computer automatically changes them to degrees and decimals, the form needed in order to accomplish the necessary calculations.

29
STANDARD TEMPERATURES (ISA)

Since much of the data contained in pilot's operating handbooks is based on ISA temperatures, there are times when we would like to know what this value is at a given altitude. One of the aviation computers contains this information in its memory. In all cases the results shown are in degrees Celsius.

With the Avstar, the only one in this group containing this program, we press **Alt/AS** and then enter the altitude for which we want the ISA (see Figure 29-1). We press **PAlt** followed by **Comp** and **T°C**. If the altitude is 5330, the airport elevation at Denver, then the result will be displayed as *4*, meaning that +4°C is the ISA for Denver.

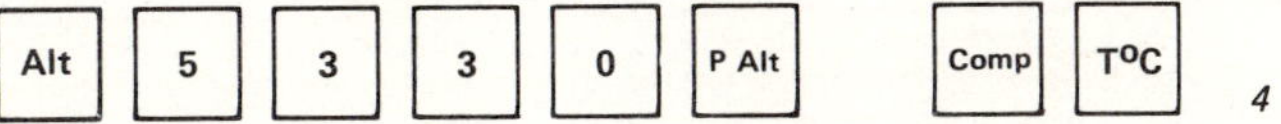

FIGURE 29-1 The Avstar is the only aviation computer containing a standard-temperature program.

30
TIMER AND STOPWATCH FUNCTIONS

Several of the aviation computers include a timer function. These timers will display the time in either increments of seconds or decimals, starting at 00:00:00 and reading up to 99 hours, 59 minutes, and 59 seconds, or they will read down from any time value entered by the pilot until 00:00:00 is reached, at which point one will emit an audible alarm.

These timer functions are extremely valuable for a variety of flight-related situations:

Time en Route With the computer attached to the appropriate converter and plugged into the cigarette lighter (in order to prevent the batteries from going dead), we can start the timing function at lift-off and keep it going throughout the flight, having a constant display of exactly how long we have been in the air. This does not prevent us from using the other computing and calculating functions whenever we want to. They are designed to have the timer continue its counting even though the computer is used for other purposes (as long as it is not turned off). At any time we can recall the time display to see how much time has passed.

Instrument Approach The count-down-plus-alarm feature is a great help during the timed portion of instrument approaches. At any time prior to reaching the fix we can set the time shown on the approach plate without actually starting the timing function. Then, at the moment at which we cross the fix, we can punch the appropriate key and the timer will start to count down. No need to continuously watch it. When it reaches zero an alarm will sound.

Determining Ground Speed without DME   We set the display on zero and press the start key when passing over a VOR or other known fix. The timer will start counting up and when we reach the next VOR, the distance to which we have determined from the chart, we stop the timer. We now know the time it took to cover the distance and it is an easy matter to compute the ground speed from it.

Fuel Burned and Fuel Remaining   As long as we know the fuel flow in terms of gallons per hour or pounds per hour, we can use the timer to determine the amount of fuel burned up to any given moment or the amount of fuel remaining. How this can best be accomplished is shown in the examples and illustrations below.

The timer function in the OC-1401 is designed to perform three different types of functions (see Figure 30-1). It can keep track of current local or Greenwich time. It can display elapsed time. And it can display the amount of fuel remaining as the flight progresses. All of these functions can be used simultaneously.

To keep track of actual time, we set the takeoff time, say, 12:00:00 into the display by entering 12.00. (two decimal points must be used to tell the computer that what we're after is a time value). The display now shows *12.00.00*. We then press **F SET** and the display changes to *12.00.01*, *12.00.02*, etc., and it will continue to do so until the computer is told to stop, or until the computer is turned off. (The timer function does not operate in the **STANDBY** mode, so if we want to keep using it for any length of time, the computer must be plugged into the cigarette-lighter outlet in the aircraft or into the house current at home or office.) We can now get rid of the display by pressing **CLR X**. The display changes to *0.00* but the clock continues to run. Now, let's say that we have run a bunch of other computations, and we want to know what time it is right now. We press **F-1 CLK** and the display comes up with the current time, and continues to count up.

To use the elapsed-time (stopwatch) function, we simply press **S/S ET** and the display will show *0.00.00* and will start to count up: *0.00.01*, *0.00.02*, etc. If we want to stop that counting, we can press **S/S ET**

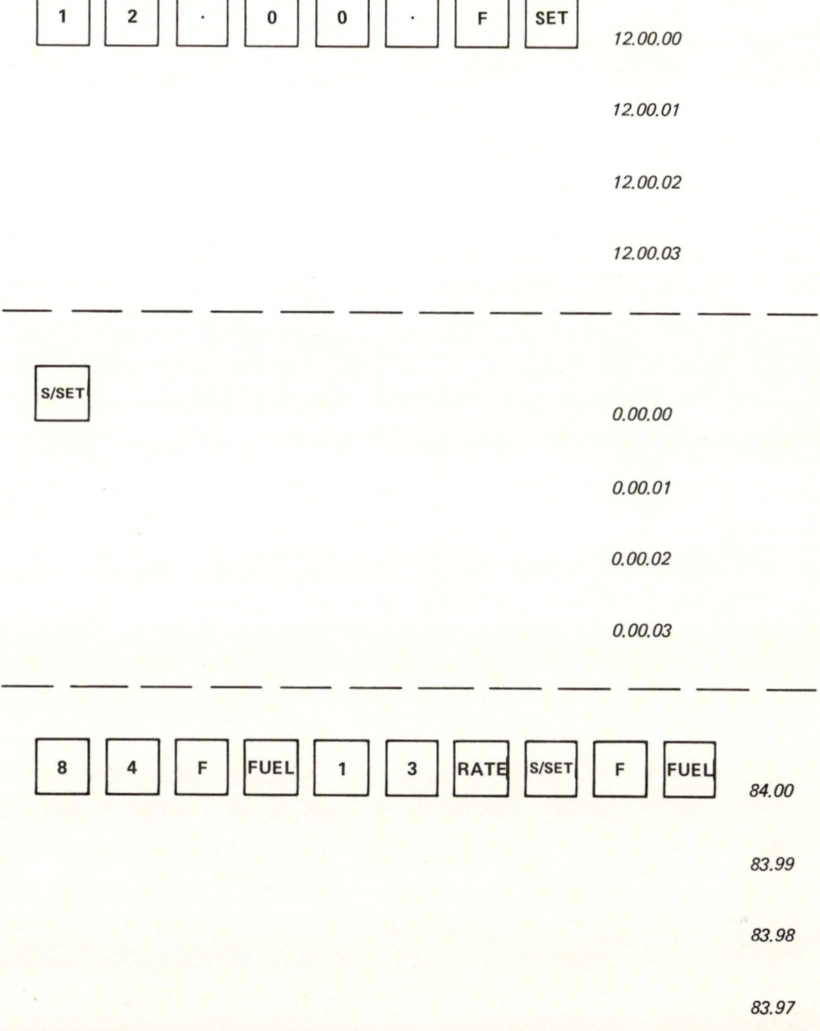

FIGURE 30-1 The OC-1401 contains a timer which can perform several functions at once.

again and it stops. (The **S/S ET** stands for *start/stop elapsed time*.) Then, if we press the same key again, it starts to count up again from where it left off. By pressing **CLR X** we can get rid of that display, but both timer functions continue to operate. At any later time, assuming the computer has not been turned off or put in **STANDBY** mode, we can press **F-1 CLK** for the actual time and **F-1 ET** for a display of the elapsed time.

The third function has to do with fuel consumed and fuel remaining. In order to perform this function, the amount of fuel at the beginning of the trip must be entered into the computer, as must the rate of fuel flow. This is accomplished, assuming 84 gallons and 13 gal/h, by pressing **8 4 F FUEL 1 3 RATE**. (Note that the computer has two keys marked **FUEL**. For the above action the *preflight* **FUEL** key must be used. It is located in the orange section of the keyboard.) Now, as soon as the engine is started, we press **S/S ET** to start the elapsed-time function. When we then press **F FUEL** (this time the **FUEL** key in the yellow *in-flight* portion of the keyboard) the amount of fuel remaining will show up in the display, and it will count down in increments of 0.01 gallon and, if running until no fuel is remaining, all kinds of crazy figures show up and an *F* starts to flash madly to tell the pilot to refuel. Throughout all this the actual time and elapsed-time functions continue to operate and can be called up by pressing the appropriate keys. Also, all the time-function displays can be cleared by pressing **CLR X** and the computer can be used for all manner of other calculations or computations without interfering with the continuing timing functions. (By the way, the *F* will start to flash just prior to fuel exhaustion, even if the computer display is used for other than fuel data.)

While the OC-1401 is not designed to count down time, meaning to count down from a preset time value to zero (something which the Navtronic does, and which can be useful during instrument approaches), there is a way to make it do that just the same. The function to use is the fuel-remaining function. Here is how that works: In order to achieve a fuel flow of 0.01 gallon per second (gal/s), the fuel flow per hour is 36 gallons. Thus, if we feed in any given amount of fuel (in this instance representing time) and enter a fuel-flow figure of 36 gal/h, the last digit in the display will change downward by 1 once every second.

Now let's assume that the approach plate stipulates that 3 minutes and 42 seconds after passing a given approach fix (assuming a certain airspeed), we have to start the final descent. A fuel flow of 36 gal/h equals a fuel flow of 0.6 gallon per minute (gal/min). So we now enter 0.6 and multiply it by 3 (see Figure 30-2). We then store the result in one of the memories and enter 0.01 and multiply it by 42. We then recall the value from memory and add it to that last result. The final

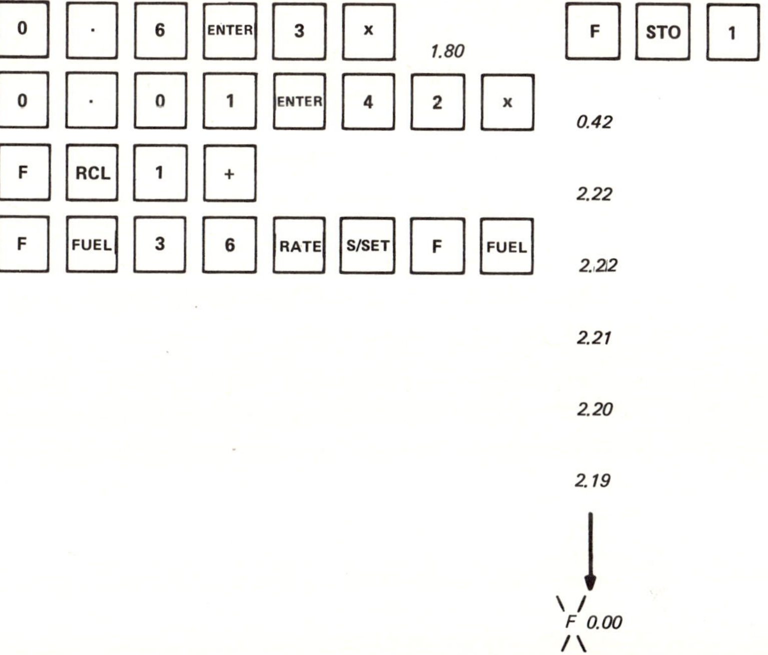

FIGURE 30-2 Though the timer function of the OC-1401 is not designed to count down time, there is a way in which that can be accomplished.

display will show *2.22*. What this means, in fact, is that 222 seconds equal 3 minutes and 42 seconds.

Having arrived at this we enter it as the amount of fuel (time) available and then enter 36 as the rate of fuel flow. All of this should be done long before we get to that final-approach fix. Then, as we pass the fix we press **S/S ET** and follow it (no hurry) by pressing **F FUEL** (in the yellow area) and the display will count seconds down to zero, at which point a flashing *F* tells us to start our descent.

The Navtronic 1701t and 1701tr have the built-in timer function. To activate the timer we start by pressing **CONV** and **T/C** which produces a display showing *00 00 00* (see Figure 30-3a). We now press **START** and the display will, at 1-second intervals, change to *00 00 01*, *00 00 02*, etc.

In order to cause the timer to count down, we first enter the time value we want (2 minutes and 6 seconds). We then press **CONV**, **T/C**, and **START** and the display will change at 1-second intervals: *00 02 06*,

00 02 05, etc., to *00 00 00* at which point an alarm will ring (see Figure 30-3*b*).

In Figure 30-4 we are again using the Navtronic 1701t or 1701tr, but this time we want to keep track of the amount of fuel burned and how much is remaining in the tanks. On takeoff we start the timer just as was described before and the display shows the time which has elapsed since lift-off. After 1 hour, 34 minutes, and 55 seconds we decide to check the fuel situation. We press **CONV** and **T/C** followed by the multiply key (**×**) followed by the fuel flow in gallons per hour (11) and the display will come up with *17.401388* which represents the amount of fuel we have used so far. We now punch the deduct key (**−**) and enter the amount of fuel which was on board when we started (84). We now punch the equal key (**=**) and the change-sign key (**+/−**)

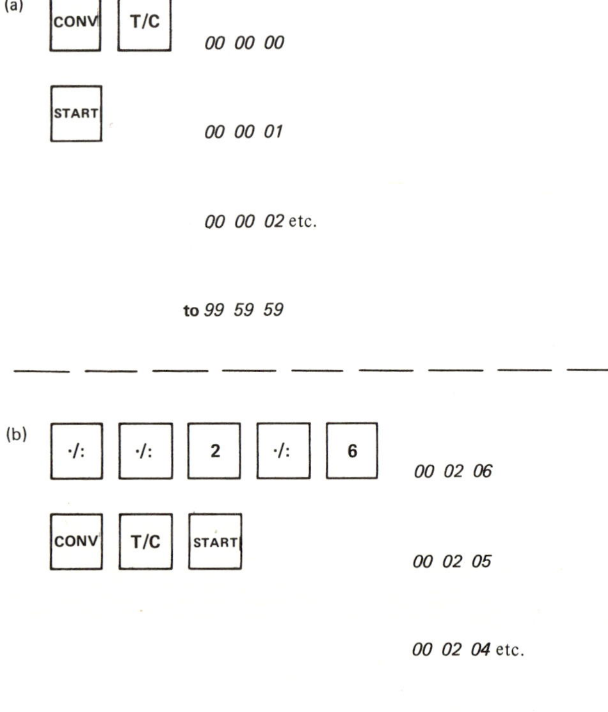

FIGURE 30-3 The Navtronic 1701t and 1701tr are equipped with a timer function which can count either up (a) or down (b).

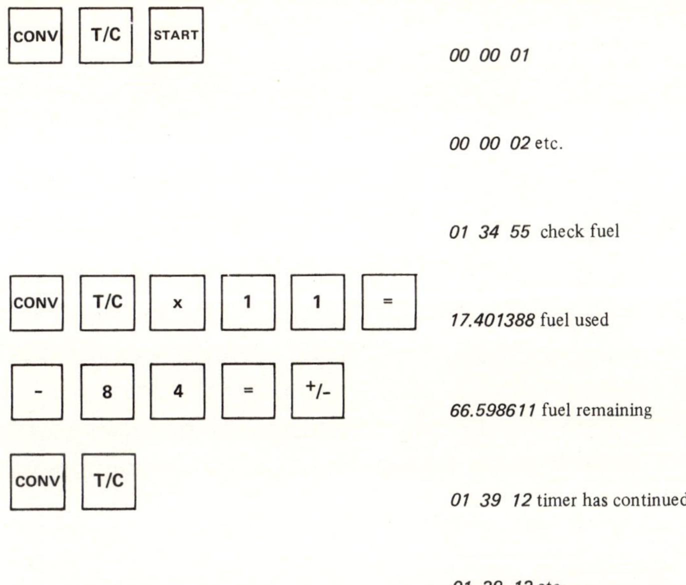

FIGURE 30-4 The timer function can be used to keep track of
fuel used and remaining.

and the display now shows **66.598611** which is the amount of fuel, in
gallons, remaining in the tanks. We now press **CONV** and **T/C** and the
display now shows **01 39 12** because the timer has continued timing
throughout this exercise and it continues to do so until shut off.

31

DME SLANT-RANGE CORRECTION

Quite logically, since all DME readings are based on the sending and return of signals from the aircraft to a ground-based navigation aid, there is always a degree of error in view of the fact that the aircraft flies horizontally, while the signals travel at an angle from the flight level to the ground, and thus cover a somewhat greater distance than the actual distance between the aircraft and the station. Also logically, this imprecision increases with the difference in altitude between the flight level of the aircraft and the elevation of the station, and it decreases with an increase in distance of the aircraft from the station. The difference between the actual distance and the slant-range distance is too small to be of consequence when we're relatively far away from the station, but at close range it is quite considerable. (As an example, an aircraft flying at 30,000 feet above a VORTAC at sea level will have a DME distance display of just under 6 nautical miles, despite the fact that the horizontal distance is 0.)

The Commodore N60 has a program which is designed to make the appropriate slant-range corrections. To use it we first identify the program to the computer by pressing **F DME** (see Figure 31-1). We then enter the speed displayed on our DME, in our illustration, 145, and identify it by pressing **SPD DME**. We follow it by the altitude differential between our current flight level and the elevation of the ground station and identify that figure by pressing **Δh**. We then tell the computer that we are 6 miles from the station by pressing **6 DIST** and then we press **CMP** and the resultant readout is *157.1548981*, the corrected ground speed.

The instruction booklet, in this instance, is again in error. It asks that the radial from the station as well as our magnetic course be entered. This is entirely unnecessary. But there is another application of the program in which that information is needed. This program corrects the speed reading not only in respect to the vertical slant range, but also to the horizontal difference which results when we're flying not directly to the station, but expect to bypass it by some miles. In the example I have assumed that the aircraft is on the 245° radial from the station, and that the course being flown is 275° magnetic. Without the computer, the resulting DME reading would be completely meaningless. To work this program we again start with **F DME 1 4 5 SPD DME** and then follow it by the radial and course information: **2 4 5 R₁ 2 7 5 CRS** (see Figure 31-1). We then enter the altitude difference between our flight level and the station just as before, and then tell the computer that we are 11 nautical miles from the station by pressing **1 1 DIST**. That is followed by **CMP** and the resultant ground-speed figure is displayed as *171.7272754*.

The TI59 (58C) also has a program which is used for DME slant-range correction, both horizontally and vertically. The program is aviation program 12. Unlike the previously described system of solving the vertical slant-range-correction problem, this computer requires that radial and course information be entered. Thus, when we are actually flying directly toward the station, the course and radial information is represented by identical figures.

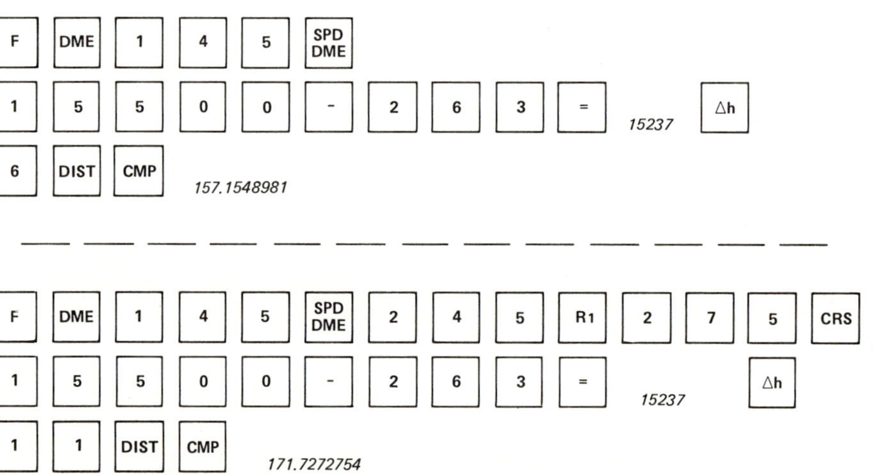

FIGURE 31-1 The N60 has a program designed to perform DME slant-range corrections.

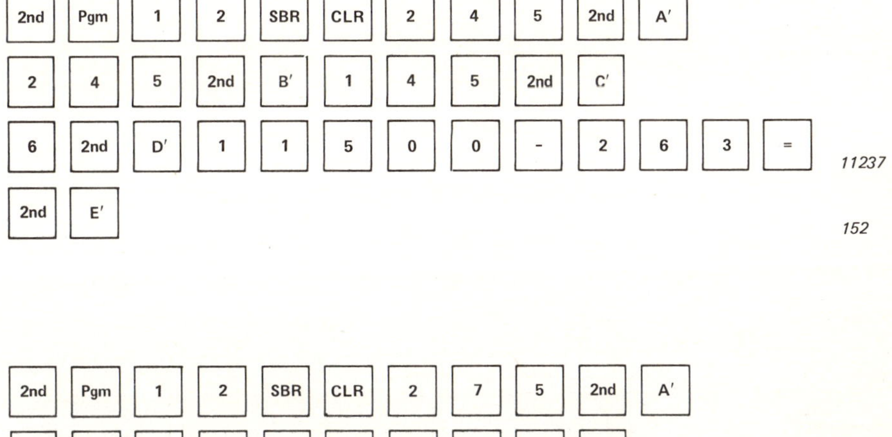

FIGURE 31-2   The TI59 uses its AV-12 program to perform DME slant-range corrections.

As whenever we use a TI aviation program, we start by identifying and then initializing the program in the previously described manner (see Figure 31-2). We then enter the course and place it into key position **A′**. We now enter the radial from the station on which the aircraft is located and place it into key position **B′**. We then enter the ground speed as presented in the DME readout, and place it into key position **C′**. Next comes the distance from the ground station as presented in the DME readout. It is placed into key position **D′**. We then deduct the station elevation from our flight altitude and place the result into key position **E′**. Without further ado, the computer will display *152*, the true ground speed as corrected for the vertical slant range.

To figure a correction for both vertical and horizontal slant range, the routine is identical, except that instead of the values for course and radial (placed into key positions **A′** and **B′**) being identical, they are different (see Figure 31-2).

32

WEIGHT AND BALANCE

This is one of those tricky problems which most of us like to avoid except when extremely critical loading of the aircraft is unavoidable. And then it is a safe bet that most of us don't remember offhand how weight and balance must be figured and what the various moments and arms are relative to our aircraft. A *moment* is the figure which represents the rotating force times its perpendicular distance from the axis of rotation. The *arm* is the distance from a so-called *datum line* located usually more or less in line with the tip of the spinner. The mathematics involved in determining weight and balance require that the length of each arm (measured in inches) be multiplied by the weight in pounds producing a result in inch-pounds. Adding up all weight figures will produce the total gross weight of the aircraft as loaded, and adding up the moment figures results in the total inch-pounds. The moment total must now be divided by the total weight and the result represents the center of gravity (CG) location in inches from the datum line. Both the total gross weight and the location of the CG must be within the allowable limits of the aircraft (as stated in the pilot's operating handbook) in order for the aircraft to be flown safely and legally.

As can be seen by the above, the problem involves pure mathematics and can therefore be solved by any standard calculator or by the calculator sections of the various aviation computers. In order to make it somewhat simpler, the Navtronic has a built-in weight-and-balance program which reduces the necessary steps to a certain degree. But even then, solving this program is, in fact, a big pain, because of the need to enter into the computer (or calculator) all those different moment/arms for fuel, oil, basic empty airplane, pilot, passengers in their different seats, and baggage and freight in its different locations. The applicable figures are available in the pilot's operating handbook and the only variables which have to be entered in addition to those figures are the weights in pounds of each item mentioned above.

Here is where some of those sophisticated (and expensive) programmable computers come in handy. With these computers the pilot can once and for all enter all of the moment/arm figures into a permanent memory and then, whenever a weight-and-balance problem must be solved, the pilot simply enters the weights into the computer and then runs the prestored program, and the computer will do the rest.

For those readers who want to know how to run such a program on a standard calculator or on the calculator portions of their aviation computers, here, in consecutive order, are the steps which must be taken.

The assumptions are these:

Item	Weight, lb	Arm, in	Moment
Basic empty weight, equipped	2000	+34.3	68634.3
Pilot and front passengers	340	+36	
Rear-seat passengers	340	+70	
Fuel (80 gal av gas)	480	+48	
Baggage	100	+95	

Now let's take it step by step:

1. Moment of basic empty airplane (68634.3) is initially stored in the memory.

2. Pilot and front-seat passenger weight (340) is now multiplied by the arm (36) to produce *12240*.

3. We now recall the first figure from memory and add it to this new total, causing the display to show *80874.3*. We now again store this figure in the memory.

4. We now multiply the weight of the rear-seat passengers (340) by that arm (70) and the result is *23800*.

5. We again recall the stored figure from memory and add it to the above, resulting in *104674.3*. And we once more store this result in the memory.

6. Now we multiply the weight of the fuel (480) by its arm, and the total is *23040*.

7. This is again added to the previous total by recalling the stored number from memory. The result: *127714.3*. And again this total is put back into the memory.

8. Now we multiply the baggage weight (100) by its arm and arrive at *9500*.

9. We once more recall the stored figure and add it to the above for a total of *137214.3*. And now this figure is once more stored in the memory.

10. We now add all the weight figures (2000 + 340 + 340 + 480 + 100) and find that the gross weight, as loaded, is *3260* pounds. If this figure is within the allowable maximum gross weight for the aircraft, we go on to the next step.

11. We now recall the total moments from the memory and divide that figure (137214.3) by the total gross weight (3260) and the center-of-gravity arm is shown as *42.090276* inches. If that falls within the allowable limits set forth in the pilot's operating handbook we are now ready to take off.

Obviously, if the particular calculator being used has no memory-storage capability, the various totals must be written down and then used each time to perform the next calculation.

33
PERSONAL PROGRAMMING

Three of the portable computers described in this part of the book are so-called *programmable* computers, meaning that in addition to the programs provided in special-purpose modules by the manufacturer, the operator can devise and write his or her own personal programs and can then ask the computer to perform those programs.

In order to appreciate what this means, let's first talk about what the term *program* means when used in conjunction with computers. In the simplest of terms it refers to a series of computations which are to be performed repeatedly in order to examine the relationship between certain fixed and variable data. In other words, to use a very simple example: The conversion of degrees Celsius to Fahrenheit involves multiplying the degrees Celsius by 9, then dividing the result by 5 and adding 32 to that sum. Thus 100°C (the boiling point of water at sea level) times 9 equals 900, divided by five equals 180, add to that 32 and the result is 212°F. In computer-program terms this would look like: **× 9 ÷ 5 + 32 =**. Once the computer has been told to run that particular program, all we have to do is to punch in the desired degrees Celsius and it will automatically repeat that series of computations, providing us with the correct answers in moments. So by punching in 31 and telling the computer to run the program, the display comes up with **87.8**, or if we punch in 0 (the freezing temperature) it comes up with **32**.

A good example of personal programming in a highly sophisticated form is what Beech has done in conjunction with the HP-41C for the Super King Air. Someone at Beech, or a computer programmer hired by Beech, has first placed all of the performance parameters applicable to that particular aircraft into the computer memory. He or she then devised a series of calculations, probably involving several hundred individual calculations, which would show the pilot exactly how long a certain flight would take under certain forecast wind conditions, and how much fuel would be consumed, assuming any of the available legal altitudes. The program was then further refined to have the computer automatically examine the results and to provide answers with respect to the fastest altitude and the altitude which would prove to be the most economical one in terms of fuel. And it will also show how much time is saved by choosing the fastest altitude, and how much fuel is saved over that by choosing the most economical altitude. (The Beech program also performs certain other functions for the Super King Air, but this is the most dramatic.) The entire program was then reduced to a special-purpose module (available from Beech, not from Hewlett-Packard) which can simply be plugged into the computer to be permanently available.

This same procedure can be followed using any type or model of aircraft and using any of the programmable computers. And it is reasonable to assume that such predesigned programs will eventually be provided by several of the manufacturers for their more popular aircraft.

In order to demonstrate what is involved, I have devised a somewhat simplified version of this type of program for the Cessna Turbo Skylane RG, using the Texas Instruments TI59. What follows is a description of each step, and the reason why it is performed.

The first step involved placing all of the fixed values into memory positions in the computer. The reason for this is that it permits using the same program for any other airplane by simply replacing the performance parameters for the Cessna Turbo Skylane RG with those applicable to any other airplane. (They could have been used in their actual form in the program, but then the program could not have been used for another airplane without changing it.)

The memory positions available for this purpose are numbered 00 through 59 and the procedure of placing data into the various memory positions is similar with all programmable computers. For the TI59 it looks like this: Value in digital form **STO 00** (or any applicable memory number). Thus storing the climb speed, 117 knots, of the Turbo Skylane RG into memory position number 01 involves these keystrokes: **1 1 7 STO 0 1**.

The memory positions used for each value are purely arbitrary. In the example I have used the memory positions 01 through 18 for the speeds, fuel flow, and rates of climb and descent, based on the pilot's operating handbook. Positions 20 and 30 are reserved for the elevation of the departure and destination airports. Positions 21 through 28 are used to enter the forecast wind components, assuming they are + (tail wind) values, while comparable − (head wind) values are stored in positions 31 to 38. Positions 41 through 49 were left blank to be used to store subtotals which are part of the program. Then, into positions 51 through 56 I have placed all acceptable altitudes up to the aircraft's maximum certificated altitude of 24,000 feet, at 4000-foot intervals. And position 59 is used for the total distance between departure and destination. As I said before, there is no particular reason for using these specific positions. Any others would have done just as well. Table 33-1 lists the contents of all 60 memories at the point at which we're ready to actually start running the program.

TABLE 33-1 DATA STORED IN MEMORY FOR THE AUTHOR'S PROGRAM TO SELECT THE BEST ALTITUDE (Aircraft: Cessna Turbo Skylane RG)

Memory	Value	Description
00	Empty	
01	117	Climb speed (all speeds in knots)
02	130	TAS @ 4,000 ft
03	137	TAS @ 8,000 ft
04	143	TAS @ 12,000 ft
05	149	TAS @ 16,000 ft
06	153	TAS @ 20,000 ft
07	158	TAS @ 24,000 ft
08	145	TAS during descent
09	600	Rate of climb (ft/min)
10	400	Rate of descent (ft/min)
11	21.3	Fuel flow, climb (all fuel in gallons)
12	12	Fuel flow @ 4,000 ft (cruise:economy)
13	11.3	Fuel flow @ 8,000 ft
14	11.1	Fuel flow @ 12,000 ft
15	11.6	Fuel flow @ 16,000 ft
16	11.8	Fuel flow @ 20,000 ft
17	11.2	Fuel flow @ 24,000 ft
18	9	Fuel flow, descent
19	Empty	

(Continued)

TABLE 33-1 (*Continued*)

Memory	Value	Description
20	2528	Departure airport elevation
21	5	Tail wind @ 3,000 ft (winds in knots)
22	8	Tail wind @ 6,000 ft
23	10	Tail wind @ 9,000 ft
24	15	Tail wind @ 12,000 ft
25	25	Tail wind @ 15,000 ft
26	30	Tail wind @ 18,000 ft
27	35	Tail wind @ 21,000 ft
28	35	Tail wind @ 24,000 ft
29	Empty	
30	900	Destination airport elevation
31	−5	Head wind @ 3,000 ft
32	−8	Head wind @ 6,000 ft
33	−10	Head wind @ 9,000 ft
34	−15	Head wind @ 12,000 ft
35	−25	Head wind @ 15,000 ft
36	−30	Head wind @ 18,000 ft
37	−35	Head wind @ 21,000 ft
38	−35	Head wind @ 24,000 ft
39	Empty	
40	Empty	
41	Empty	
42	Empty	
43	Empty	
44	Empty	
45	Empty	
46	Empty	
47	Empty	
48	Empty	
49	Empty	
50	Empty	
51	4000	Altitude
52	8000	Altitude
53	12000	Altitude
54	16000	Altitude
55	20000	Altitude
56	24000	Altitude
57	Empty	
58	Empty	
59	750	Total trip distance

At this point it would be a good idea to record the data stored in the memories in some reusable form. With the TI59 and the HP-41C, assuming the latter is equipped with the optional card reader, this can be done by causing the computer to "write" the stored information on one of these magnetic cards. Then, at any time in the future, we can simply run that card through the computer, and all the data is returned to its applicable memory positions.

The program itself, taken step by step, looks like this:

Display	Keystroke	Comment
0	**2nd**	Changes the meaning of the next key to be used from **CE** (clear entire computer) to **CP** (clear previously entered programs)
0	**CP**	
0	**LRN**	This key tells the computer to learn (**LRN**) the program which we are about to key in. The display now changes to *000 00* to show that no program step has as yet been executed
000 00	**RCL**	This key is used to recall (**RCL**) a value for a given memory position
001 00	**56**	This identifies the memory position in which the value is stored which we want to use in the first calculation. In this case, the cruising altitude of 24,000 ft
002 00	—	The minus (—) key is pressed because the next value is to be subtracted from that cruise altitude
003 00	**RCL**	
004 00	**20**	We have now recalled the airport elevation at our departure point and told the computer that it is to deduct it from the cruise altitude
005 00	=	The computer performs the calculation. Note that it does not display the result. All it displays is the number of steps which have been performed in the program so far
006 00	÷	We now tell the computer that we want to divide the result of the previous calculations, representing the altitude change to be accomplished during climb, by the rate of climb in order to determine the amount of time spent in climb
007 00	**RCL**	
008 00	**09**	The rate of climb
009 00	=	The computer now has calculated the time spent in climb. Since we will need this figure again later on in the program, we now tell the computer to store it in one of the empty memory positions

(Continued)

Display	Keystroke	Comment
010	**STO**	The command to store (**STO**) information
011	**40**	The memory position
012	**×**	Even though the previous total has been stored in memory 40, it is also still active in the computer and can thus be used for the next calculation which calls for multiplication by the expected average ground speed throughout the climb. Since this ground speed is the result of subtracting the average head wind component (or adding the average tail wind component) from (to) the climb speed, the next computation must be placed in parentheses
013 00	**(**	
014 00	**RCL**	
015 00	**01**	
016 00	**+**	
017 00	**RCL**	
018 00	**34**	
019 00	**)**	What has been done here is to recall the climb speed (01) and subtract from it the head wind at 12,000 ft (31) which can be assumed to represent the average between takeoff and cruising altitude. The reason why a **+** (plus) rather than a **−** (minus) sign is used, is that the wind component stored in the memory is expressed as -15
020	**=**	We now tell the computer to complete the calculation started with step 013, namely, to multiply the result of the above calculation by the time spent in climb. (The step number appearing opposite a given keystroke is actually applicable to the previous keystroke, as a keystroke must be made in order for the display to change to the next number.) But even though we can't see the result which is hidden somewhere in the computer, logic tells us that by multiplying time in terms of minutes (resulting from dividing the altitude change by vertical speed in feet per minute) by speed in terms of so many units per hour (knots or miles per hour, in this case knots) the result must be divided by 60 in order to be correct in terms of the horizontal distance covered during climb. So that is then the next step
021 00	**÷**	
022 00	**6**	
023 00	**0**	
024 00	**=**	The computer now knows the distance traveled during climb. This is another figure which we'll need later again in the program, so we store it in another memory position

Display	Keystroke	Comment
025	STO	
026	41	Since we do not want to use this value in our next calculation, we have to clear it out of the active position in the computer
027	CLR	Clear (**CLR**)
		Now, in order to determine the amount of fuel used while climbing, we want to multiply the time spent in climb by the fuel flow at the full-power setting used during the climb
028 00	RCL	
029 00	40	
030 00	x	
031 00	RCL	
032 00	11	Climb fuel flow
033 00	=	Again the result is the product of multiplying time in minutes by fuel flow in gallons per hour, and it must be divided by 60 in order to be correct
034 00	÷	
035 00	6	
036 00	0	
037 00	=	We now store that value for future use
038 00	STO	
039 00	42	
040 00	CLR	
041 00	RCL	
042 00	56	Now, in order to repeat the entire process for the descent portion of the flight, we again recall the cruising altitude
043 00	−	
044 00	RCL	
045 00	30	Destination airport elevation
046 00	=	
047 00	÷	
048 00	RCL	
049 00	10	Rate of descent
050 00	=	Time spent in descent is again stored for future use
051 00	STO	
052 00	43	
053 00	x	With that value now stored but also still in the active portion of the computer, we multiply the ground speed which is the sum of the TAS and the wind component
054 00	(	
055 00	RCL	
056 00	08	
057 00	+	

(Continued)

Display	Keystroke	Comment
058 00	RCL	
059 00	34	
060 00	)	
061 00	=	
062 00	÷	
063 00	6	
064 00	0	
065 00	=	The distance covered in descent is again stored
066 00	STO	
067 00	44	
068 00	CLR	
069 00	RCL	
070 00	43	Now, in order to determine the amount of fuel used during descent, we recall the time spent in descent
071 00	x	We multiply it
072 00	RCL	
073 00	18	by the fuel flow during descent
074 00	=	
075 00	÷	
076 00	6	
077 00	0	
078 00	=	
079 00	STO	The result is also stored
080 00	45	
081 00	CLR	
082 00	RCL	We now call up the total trip distance and deduct from it the distances covered during climb and descent
083 00	59	
084 00	−	
085 00	RCL	
086 00	44	
087 00	−	
088 00	RCL	
089 00	41	Distance covered during cruise
090 00	=	This must now be divided by the ground speed, represented by the TAS plus the wind component
091 00	÷	
092 00	(	
093 00	RCL	
094 00	07	
095 00	+	
096 00	RCL	
097 00	38	
098 00	)	
099 00	=	Time spent in cruise
100 00	STO	Store for future use

Display	Keystroke	Comment
101 00	**47**	
102 00	**×**	Multiply that value by
103 00	**RCL**	
104 00	**17**	the fuel flow at 55% at 24,000 ft
105 00	**=**	Fuel used in cruise
106 00	**STO**	Store
107 00	**48**	
108 00	**CLR**	
109 00	**RCL**	We now recall the time spent in cruise and add to it the time spent in climb and descent to arrive at the total time for the trip. Climb and descent times must again be divided by 60 in order to represent hours rather than minutes
110 00	**47**	
111 00	**+**	
112 00	**(**	
113 00	**RCL**	
114 00	**40**	
115 00	**÷**	
116 00	**6**	
117 00	**0**	
118 00	**)**	
119 00	**=**	
120 00	**+**	
121 00	**(**	
122 00	**RCL**	
123 00	**43**	
124 00	**÷**	
125 00	**6**	
126 00	**0**	
127 00	**)**	
128 00	**=**	
129 00	**STO**	This is now the time spent in cruise
130 00	**58**	Store
131 00	**CLR**	
132 00	**RCL**	We now add the fuel totals for climb, descent, and cruise
133 00	**48**	
134 00	**+**	
135 00	**RCL**	
136 00	**42**	
137 00	**+**	
138 00	**RCL**	
139 00	**45**	
140 00	**=**	Total fuel used during the trip
141 00	**STO**	Store

(Continued)

Display	Keystroke	Comment
142 00	**00**	
143 00	**R/S**	Tells the computer that this is the end of the program
144 00	**LRN**	Tells the computer to stop learning
0		
		We are now ready to run the program
0	**RST**	Restart
0	**R/S**	Run
C		The computer takes a few seconds to run the program. The display then comes up with (set for 4 decimals):
72.6691		representing the fuel used for the entire trip. Now, in order to find out the time spent traveling, we recall the value stored previously
	RCL	
	58	
6.0172		The total time in hours and decimals of hours. We now convert that to hours, minutes, and seconds
	INV	
	2nd	
	D.MS	
6.0102		6 h, 1 min, 2 s
		Now, if, for some reason, we want to break those time and fuel figures down into climb, cruise, and descent, all the data is still stored in its different memory positions
		To call up fuel used in climb:
	RCL	
	42	
14.0761		Climb fuel
		To find time in climb:
	RCL	
	40	
35.7867	**INV**	
	2nd	
	D.MS	
35.4712		35 min, 47 s
		To find fuel used in cruise:
	RCL	
	00	
49.9330		Cruise fuel
		To find the time spent at 24,000 ft:
	RCL	
	47	
4.4583	**INV**	
	2nd	
	D.MS	
4.2806		4 h, 28 min, 6 s
		To find fuel used in descent:

Display	Keystroke	Comment
	RCL	
	45	
8.66		Descent fuel
		To find time spent descending:
	RCL	
	43	
57.7500		
	INV	
	2nd	
	D.MS	
57.4500		57 min, 45 s
		Then, in order to know how far from our destination we have to start our descent in order to arrive with no altitude to spare, we can call up the distance to descend:
	RCL	
	44	
139.5625		meaning that 139.5 mi prior to arrival at our destination we have to set up a 400 ft/min rate of descent, retaining a TAS of 150 kn

Now, before turning the computer off, meaning that the entire program is still stored within its innards, we store the program in its entirety on a magnetic card. With the program thus permanently available, we can rerun it at any time, using different wind information, different altitudes, or different aircraft parameters. All that needs to be done is to first reenter all of the memory information by feeding the magnetic card which permanently contains that information through the computer. Then, in all positions in which new data needs to be entered, we simply place that new data into the appropriate memory positions, replacing the old data. Then, by feeding the magnetic card with the program through the computer, we restore the entire program and when we tell it to go ahead and run the program, it will come up with new results based on the new data. In a similar manner the program can be run over and over again with the same data, but using different altitudes and the wind information relative to those altitudes, and can thus arrive at a determination showing us which altitude would be the most economical, and which the fastest. Then, by comparing the results, we can figure out whether the savings in terms of fuel are worth the extra time spent en route by selecting the most economical altitude.

As an example, by running the same program for three typical altitudes (4000, 12,000, and 24,000 feet) with the appropriate wind components, the results would look like this:

Altitude, ft	No wind	Tail wind	Head wind
4,000 time:	5:47:10	5:34:19	6:01:04
fuel:	69.4276	66.8569	72.2050
12,000 time:	5:17:33	4:48:26	5:54:08
fuel:	57.2240	51.8316	63.9944
24,000 time:	4:58:51	4:14:21	6:08:41
fuel:	59.6917	51.3845	72.7263

The underlined figures show the fastest and the most economical altitude for each of the three wind conditions.

As you keep working with these computers you'll find that they don't always come up with the identical result, right to the last decimal. That tends to be annoying at first, but in fact the discrepancies are sufficiently minor as to be of no consequence.

Programming is a very personal activity. Different individuals will tend to approach the solution to a problem in different ways. There is no hard and fast rule, as long as it is understood that each program, in order to work, must contain all the commands for the necessary calculations in the correct order. Once one gets the hang of programming it's a lot of fun and it certainly can help to reduce the time involved in performing complicated mathematical computations.

To prove the personal aspect of this type of programming, I asked my son (Peter Garrison of Melmoth fame) to create an alternative but similar program on his TI59 and, since he also has the optional printer, to print it out. The result is actually a better program because, instead of having to calculate wind components to be encountered at altitude, based on winds-aloft forecasts, this program incorporates the wind-vector program contained in the aviation module, permitting the use of the actual forecast winds.

Figure 33-1*a* is a printout of the data which is stored in the various memories in the computer:

Memory location	Data stored
05	Magnetic variation
09	True course
19, 20	Speeds during climb and descent
21–30	Fuel flows
31–38	Winds
40, 41	Rates of climb and descent
42, 43	Airport elevations
47	Total distance
51–58	True airspeeds

All those other unidentified numbers in Figure 33-1*a* are meaningless to us. They are simply values used by the computer when conducting its computations.

Figure 33-1*b* shows the program divided into individual sections, each one identified by a label (**LBL**). The first nine sections (lines **000** to **142**) are designed to place the different values already entered into the memories into a form which the computer can subsequently digest. This process involves a total of 142 steps. Steps number **143** through **291** represent the actual program which determines the time and fuel data for all potential flight levels. The next section, steps **292** through **298**, performs some internal rearrangement of data within the computer. The last section, steps **299** through **312**, calls up the wind-component program from the Aviation Module (Program 07) which is utilized by the computer in order to determine the actual head- or tail wind components based on the true course, the magnetic variation, and the flight altitude.

In preparing his version of the program Peter assumed a true course of 157° and a magnetic variation of +6°. He further assumed the following forecast winds aloft:

Altitude, ft	Direction	Velocity, kn	Entered into the computer as
3,000	030°	10	3.03010
6,000	060°	12	6.06012
9,000	060°	16	9.06016
12,000	090°	20	12.09020
15,000	090°	22	15.09022
18,000	060°	25	18.06025
21,000	060°	33	21.06033
24,000	030°	30	24.03030

Peter used the same performance parameters for the Cessna Turbo Skylane RG, and the same departure and destination elevation. He then picked flight levels which coincide with the altitudes for which winds are forecast, as that seemed more logical.

The final program involves a total of 312 steps and the results for the selected altitudes can either be printed out (if the optional printer is available) or they can be called up one by one in the following manner: Press altitude followed by **2nd A′** to obtain the time from takeoff to landing in hours and decimals of hours (**6 0 0 0 2nd A′** and the display shows *5.61*). Then press **2nd B′** and the display will change to the amount of fuel required (**2nd B′** and the display shows *63.98*). This procedure can then be repeated for all other altitudes to produce information as to which altitude is the fastest and which is the most economical. Table 33-2 shows the computed time and fuel figures for all seven altitudes from 6000 to 24,000 feet, proving that under these conditions the highest altitude is not only the fastest but also the most economical.

Now, while we're at it, we may want to figure out the return flight, assuming that there is no change in the winds. (Since the wind data is stored in memories 31 through 38, any one or all can easily be changed to new data by simply punching in the new values followed by **STO** and the specific memory number.)

In order to calculate the return flight, the only figure that needs to be changed is the true course which is stored in memory number 09. So, with the complete program in the computer, we punch in the

**TABLE 33-2 ALTERNATIVE PROGRAM OUTGOING
FLIGHT TIME AND FUEL FIGURES FOR SEVEN
FLIGHT LEVELS (Aircraft: Cessna Turbo Skylane RG)**

Altitude, ft	ETE	Fuel, gal	ETE, hr:min:s
TRUE COURSE: 157°		**DISTANCE: 750 nmi**	
6,000	5.61	63.98	5:36:36
9,000	5.47	60.54	5:28:12
12,000	5.69	62.25	5:41:24
15,000	5.58	62.62	5:34:48
18,000	5.18	58.51	5:10:48
21,000	5.11	57.36	5:06:36
24,000	4.77	51.39	4:46:12

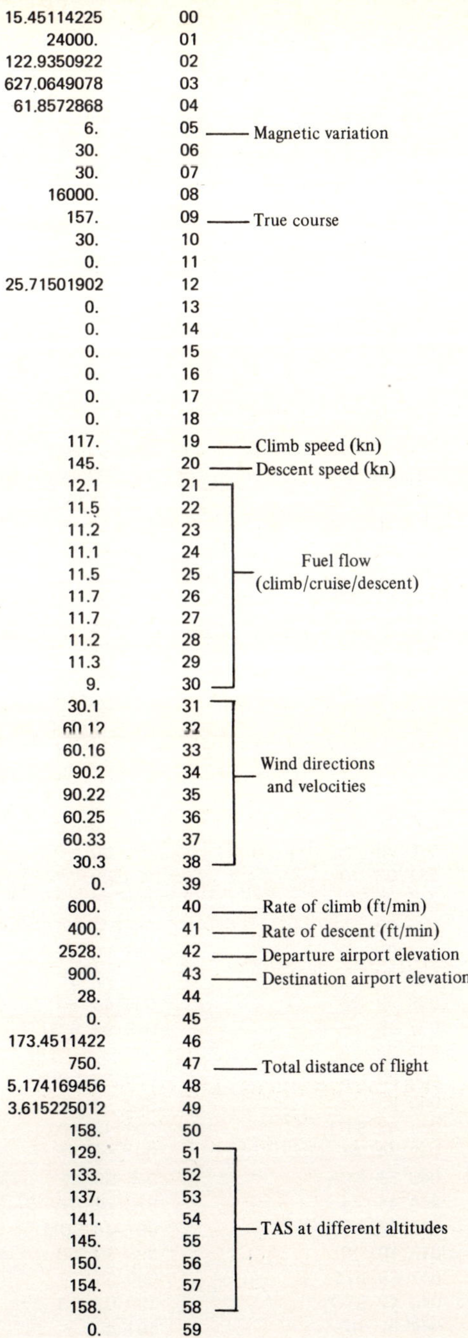

FIGURE 33-1*a*   Printout of flight data from the TI59 computer memory for the alternative program written for the Cessna Turbo Skylane RG.

```
000 76 LBL        053 01   1        105 95   =
001 71 SBR        054 00   0        106 42 STO
002 42 STO        055 00   0        107 44  44
003 50  50        056 49 PRD        108 01   1
004 59 INT        057 10  10        109 00   0
005 22 INV        058 43 RCL        110 00   0
006 44 SUM        059 10  10        111 49 PRD
007 50  50        060 42 STO        112 50  50
008 55   ÷        061 06  06        113 43 RCL
009 03   3        062 92 RTN        114 50  50
010 95   =                          115 72 ST✶
011 85   +                          116 44  44
012 03   3        063 76 LBL        117 92 RTN
013 00   0        064 61 GTO
014 95   =        065 42 STO
015 42 STO        066 50  50        118 76 LBL
016 44  44        067 59 INT        119 18  C'
017 01   1        068 22 INV        120 42 STO
018 00   0        069 44 SUM        121 09  09
019 00   0        070 50  50        122 92 RTN
020 00   0        071 55   ÷
021 49 PRD        072 03   3        123 76 LBL
022 50  50        073 95   =        124 14   D
023 43 RCL        074 85   +        125 42 STO
024 50  50        075 05   5        126 47  47
025 72 ST✶        076 00   0        127 92 RTN
026 44  44        077 95   =
027 92 RTN        078 42 STO        128 76 LBL
                  079 44  44        129 15   E
                  080 01   1        130 42 STO
028 76 LBL        081 00   0        131 42  42
029 48 EXC        082 00   0        132 92 RTN
030 55   ÷        083 00   0
031 03   3        084 49 PRD        133 76 LBL
032 00   0        085 50  50        134 10  E'
033 00   0        086 43 RCL        135 42 STO
034 00   0        087 50  50        136 43  43
035 95   =        088 72 ST✶        137 92 RTN
036 59 INT        089 44  44
037 85   +        090 92 RTN        138 76 LBL
038 03   3                          139 19  D'
039 00   0                          140 42 STO
040 95   =        090 76 LBL        141 05  05
041 42 STO        092 52 EE         142 92 RTN
042 44  44        093 42 STO
043 73 RC✶        094 50  50        143 76 LBL
044 44  44        095 59 INT        144 16  R'
045 42 STO        096 22 INV        145 99 PRT
046 10  10        097 44 SUM        146 58 FIX
047 59 INT        098 50  50        147 02  02
048 42 STO        099 55   ÷        148 42 STO
049 07  07        100 03   3        149 01  01
050 22 INV        101 95   =        150 65   X
051 44 SUM        102 85   +        151 02   2
052 10  10        103 02   2        152 55   +
                  104 00   0
```

FIGURE 33-1*b* Printout of the alternative program on the
TI59, written for the Cessna Turbo Skylane RG.

```
153 03    3
154 95    =
155 42 STO
156 08   08
157 71 SBR
158 48 EXC
159 71 SBR
160 36 PGM
161 43 RCL
162 01   01
163 75    −
164 43 RCL
165 42   42
166 95    =
167 55    ÷
168 43 RCL
169 40   40
170 95    =
171 55    ÷
172 06    6
173 00    0
174 95    =
175 42 STO
176 48   48
177 65    X
178 53    (
179 43 RCL
180 46   46
181 85    +
182 43 RCL
183 19   19
184 54    )
185 95    =
186 42 STO
187 02   02
188 43 RCL
189 48   48
190 65    X
191 43 RCL
192 29   29
193 95    =
194 42 STO
195 04   04
196 43 RCL
197 20   20
198 44 SUM
199 46   46
200 43 RCL
201 01   01
202 75    −
203 43 RCL
204 43   43
205 95    =
206 55    ÷
207 43 RCL
208 41   41
209 95    =
210 55    ÷
211 06    6
212 00    0
213 95    =
214 42 STO
215 03   03
216 44 SUM
217 48   48
218 65    X
219 43 RCL
220 46   46
221 95    =
222 44 SUM
223 02   02
224 43 RCL
225 03   03
226 65    X
227 43 RCL
228 30   30
229 95    =
230 44 SUM
231 04   04
232 43 RCL
233 47   47
234 75    −
235 43 RCL
236 02   02
237 95    =
238 42 STO
239 03   03
240 43 RCL
241 01   01
242 71 SBR
243 48 EXC
244 71 SBR
245 36 PGM
246 43 RCL
247 01   01
248 55    ÷
249 03    3
250 00    0
251 00    0
252 00    0
253 95    =
254 85    +
255 05    5
256 00    0
257 95    =
258 42 STO
259 44   44
260 73 RC*
261 44   44
262 44 SUM
263 46   46
264 43 RCL
265 03   03
266 55    ÷
267 43 RCL
268 46   46
269 95    =
270 44 SUM
271 48   48
272 42 STO
273 49   49
274 03    3
275 00    0
276 22 INV
277 44 SUM
278 44   44
279 73 RC*
280 44   44
281 65    X
282 43 RCL
283 49   49
284 95    =
285 44 SUM
286 04   04
287 43 RCL
288 48   48
289 99 PRT
290 92 RTN
291 91 R/S

292 76 LBL
293 17 B′
294 43 RCL
295 04   04
296 99 PRT
297 98 ADV
298 92 RTN

299 76 LBL
300 36 PGM
301 22 INV
302 86 STF
303 02   02
304 36 PGM
305 07   07
306 11    A
307 42 STO
308 46   46
309 92 RTN
310 00    0
311 00    0
312 00    0
```

reciprocal of 157 which is 337, followed by **STO 0 9**, and now the computer is ready to provide the time and fuel information for the return trip. Table 33-3 shows the results, showing that this time the shortest ETE can be achieved at 15,000 feet, while 12,000 feet will produce the best fuel economy.

The entire program is recorded on two magnetic cards, including all of the aircraft performance parameters, and it can be put back into the computer time and again by simply feeding the magnetic cards through it. It is thus permanently available and can be used prior to each trip, requiring only those simple corrections which are applicable to the proposed flight: Departure-airport elevation; winds aloft; true course; total distance; destination-airport elevation.

But what about FBOs or corporate flight departments which operate a bunch of different types of aircraft? The program can equally well be used for any other aircraft. All that needs to be done is to change the performance figures for the aircraft in the applicable memory positions. And, if these other aircraft are used frequently, the new parameters can be recorded on a separate magnetic card and then fed into the computer whenever the particular aircraft is to be used for a proposed flight. And the program is not limited to piston-powered airplanes. It can just as well be used for turbine aircraft. To show what happens, I replaced the performance data for the Turbo Skylane RG with performance data for the Cessna 441 Conquest turboprop. Table 33-4 shows what happens in terms of time and fuel,

TABLE 33-3 ALTERNATIVE PROGRAM RETURN FLIGHT TIME AND FUEL FIGURES FOR SEVEN FLIGHT LEVELS (Aircraft: Cessna Turbo Skylane RG)

Altitude, ft	ETE	Fuel, gal	ETE, h:min:s
TRUE COURSE: 337°		**DISTANCE: 750 nmi**	
6,000	5.65	64.45	5:39:00
9,000	5.49	60.78	5:29:24
12,000	5.05	55.18	5:03:00
15,000	4.95	55.34	4:57:00
18,000	5.06	57.06	5:03:36
21,000	4.97	55.69	4:58:12
24,000	5.24	56.57	5:14:24

**TABLE 33-4   ALTERNATIVE PROGRAM OUTGOING
AND RETURN FLIGHT TIME AND FUEL FIGURES FOR
SEVEN FLIGHT LEVELS (Aircraft: Cessna 441
Conquest Turboprop)**

Altitude, ft	ETE	Fuel, lb	ETE, h:min:s
TRUE COURSE: 157°		**DISTANCE: 750 nmi**	
6,000	3.52	1957.12	3:31:12
9,000	3.48	1891.92	3:28:48
12,000	3.63	1814.83	3:37:48
15,000	3.57	1642.20	3:34:12
18,000	3.39	1505.91	3:23:24
21,000	3.35	1403.12	3:21:00
24,000	3.16	1247.38	3:09:36
TRUE COURSE: 337°		**DISTANCE: 750 nmi**	
6,000	3.53	1962.86	3:31:48
9,000	3.49	1896.86	3:29:24
12,000	3.34	1667.16	3:20:24
15,000	3.27	1504.20	3:16:12
18,000	3.37	1494.79	3:22:12
21,000	3.33	1391.38	3:19:48
24,000	3.47	1375.78	3:28:12

using that airplane over the same 750-nautical-mile distance. Both true courses, 157° and 337°, were used in the computations, and the winds were assumed to have remained as before. As can be seen from the results, when flying the 157° course the aircraft does best at the highest altitude, both in terms of time as well as fuel. When going the other way, the fastest altitude proves to be 15,000 feet, while the most economical altitude is again the highest. In this case, adding 12 minutes to the ETE will result in a fuel saving of 128.42 pounds of fuel (all fuel figures used here are in pounds).

Other owners of the TI59 or the HP-41C (with its optional card reader) may want to experiment and come up with still a different way in which to solve these altitude-selection problems. Once such a program has been designed and recorded for future use, it will, in the long run, prove to be a valuable tool in reducing the amount of fuel being consumed.

GLOSSARY OF TERMS AND ABBREVIATIONS

A

ABQ Albuquerque, New Mexico.
ADI Attitude director indicator.
ALS Alamosa, Colorado.
arc cosine The inverse function of cosine.
arc sine The inverse function of sine.
arc tangent The inverse function of tangent.
arithmetic logic Performing calculations in normal succession, such as: 1
   plus 1 equals 2.

C

°C Degrees Celsius.
CAS Calibrated airspeed.
CDI Course deviation indicator.
CDU Control display unit.
COS Colorado Springs, Colorado.
cosine The function that, for an acute angle, is the ratio between the side
   adjacent to the angle, when it is considered part of a right triangle and the
   hypotenuse.
course The planned direction of flight in the horizontal plane.
CRT Cathode-ray tube.

D

DEN Denver, Colorado.
density altitude Pressure altitude corrected for prevailing temperature
   conditions.
DH Decision height.

diurnal shift   The fluctuation in the propagation of VLF signals which are affected by day, night, time of the year, and other predictable phenomena.
DME   Distance measuring equipment.

E

EADI   Electronic attitude director indicator.
ECDI   Electronic course deviation indicator.
EHSI   Electronic horizontal situation indicator.
ETA   Estimated time of arrival.
ETD   Estimated time of departure.
ETE   Estimated time en route.

F

°F   Degrees Fahrenheit.
FAA   Federal Aviation Administration.
FCS   Flight control system.
FL   Flight level (FL 410 = 41,000 feet).
ft/min   Feet per minute (climb or descent).

G

G-force   Force of gravity.
GIGO   Garbage in, garbage out (computer lingo).
gal/h   Gallons per hour.

H

heading   The direction in which the nose of the airplane is pointing in flight.
Hg   Mercury.
HSI   Horizontal situation indicator.
HUD   Heads-up display.

I

IAF   Initial approach fix.
IAS   Indicated airspeed.
IFCS   Integrated flight control system.
IFR   Instrument flight rules.

ILS   Instrument landing system.
inHg   Inches of mercury (the means of measuring barometric pressure).
INS   Inertial navigation system.

K

knots (kn)   Nautical miles per hour.

L

latitude   The east-west demarcation lines used in geography, measured in
degrees:minutes:seconds.
LCD   Liquid crystal display.
LED   Light-emitting diode display.
logarithm   The exponent that indicates the power to which a number is
raised to produce a given number.
longitude   The north-south demarcation lines used in geography. All longitudes dissect the north and south pole. Measured in degrees:minutes:
seconds.
LVS   Las Vegas, New Mexico.

M

Mach   The speed of sound.
MAP   Missed approach point.
mi/h   Statute miles per hour.
msl   Mean sea level.

N

natural logarithm   Logarithm with e (2.7182818+) as a base.
NCU   Navigation computer unit.
NDB   Nondirectional beacon.
nicad   Nickel-cadmium (batteries).
nmi   Nautical mile(s).
Nonvolatile   Refers to computer memories which retain input information
even when the computer is turned off.

O

OAT   Outside air temperature.

P

PNI   Pictorial navigation indicator.
Polish logic   Calculations performed in an inverse manner, such as: 1 ENTER 1 plus produces 2.
lb/h   Pounds per hour.
PUB   Pueblo, Colorado.

R

radian   A unit of plane angular measurement that is equal to the angle at the center of a circle, subtended by an arc equal in length to the radius.
RNAV   Area navigation.
ROC   Rate of climb.
root   A quantity taken an indicated number of times as an equal factor.

S

SAF   Santa Fe, New Mexico.
scientific notation   Numbers expressed as products consisting of a number between 1 and 10, multiplied by an appropriate power of 10.
SID   Standard instrument departure.
sine   The function that for an acute angle is the ratio between the side opposite the angle when it is considered part of a right triangle and the hypotenuse.
sm   Statute mile(s).
square   To multiply a number by itself.
square root   The opposite of square (the square root of 9 is 3).
STAR   Standard terminal arrival route.

T

tangent   A function that for an acute angle is the ratio between the side opposite the angle when it is considered a right triangle and the side adjacent.
TAS   True airspeed.
TAS   Taos, New Mexico.

U

upper winds   Winds aloft.

V

VFR Visual flight rules.
VHF Very high frequency.
VLF Very low frequency.
VNAV Vertical navigation.
volatile Refers to computers which lose all stored data when the computer is turned off.
VOR Very high frequency omnidirectional radio range.
VOR/DME A VOR associated with distance measuring capability.
VORTAC VOR/TACAN, for all practical purposes the same as a VOR/DME.
VSI Vertical speed indicator.

W

winds aloft Wind direction and velocity at 3000-foot intervals (3000, 6000, 9000 feet msl, etc.).
WP Waypoint.

Z

Zulu time Greenwich mean time.

ABOUT THE AUTHOR

PAUL GARRISON has published more than a dozen books on aviation. As a pilot with 30 years experience, he has a twin-engine rating although he prefers flying single-engine planes. He devotes most of his time to writing about flying and has published 300 articles. One of his books, *Inside Private Aviation*, was chosen as a Flying Book Club selection.